THE FABRIC OF MOBILE SERVICES

THE FABRIC OF MOBILE SERVICES

Software Paradigms and Business Demands

Shoshana Loeb

Benjamin Falchuk

Euthimios Panagos

WILEY

A JOHN WILEY & SONS, INC., PUBLICATION

Published by John Wiley & Sons, Inc., Hoboken, New Jersey.
Published simultaneously in Canada.

Library of Congress Cataloging-in-Publication Data:

Loeb, Shoshana.
The fabric of mobile services: software paradigms and business demands / Shoshana Loeb, Benjamin Falchuk, Euthimios Panagos.
p. cm.
Includes bibliographical references and index.
ISBN 978-0-470-27799-7 (cloth)
1. Wireless Communication systems. 2. Mobile Communication Systems. 3. Computer Software–Development. I. Falchuk, Benjamin. II. Panagos, Euthimios. III. Title.
TK6570.M6L65 2009
658.8′72–dc22 2009004207

Printed in the United States of America.

10 9 8 7 6 5 4 3 2 1

To our families

–Shoshana, Ben, Thimios

CONTENTS

PREFACE

The Fabric of Mobile Services addresses a topic that is hot today and is likely to remain hot for many years to come. In contemplating the book written by Shoshi, Thimios, and Ben, I find the subject that the authors have tackled to be very complex. This is true in terms of the underlying technologies – communications, computing, storage, software, and user interfaces; in the intricacies of operations; in the search for viable business models; and in the vast choice of applications and services that can profoundly benefit and influence end-users. The complexity comes from many factors, two of which stand out. The first is that "mobile services" are recent phenomena that involve many components. As the services are launched we do not clearly understand what new uses and patterns will emerge over time. While it is possible to understand how an isolated service can improve the way we do things today, we have not yet experienced and discovered what their full impact will be. The space to contemplate is very, very large and for the most part unexplored. The second is the tangle of timescales. Application and technology advances occur in time frames that are as short as 6 to 12 months. The fast pace of infrastructure build-out and refresh happens on three- to five-year cycles. Regulatory changes and capital formation for national deployments require decades and are increasingly difficult to align with user demand and technology evolution. The creation of human capital and changes in societal habits that drive deep adoption and can explore the full benefits of "mobile services" may require generations. This book establishes the broad framework in which one can think about the subject. It covers the range from practical near-term considerations to the driving issues of long-term evolution about which we can only speculate today.

There are several facts to keep in mind. Since the early 1990s, starting almost from nothing, a set of global wireless networks was created with stunning speed and equally stunning commitment of capital. These networks and the underlying services reach, to date, almost three billion people on the planet. The revenues and profits from these networks now account for well over half the "business" activity from service providers. At the same time, overlapping the deployment of wireless networks, the internet has also covered the world and a new generation of users thinks of it as part of their everyday experience. Again major new usage patterns have emerged as have new business models. The two events – high-speed wireless and the internet, are running into each other. "Mobile services" is the most obvious outcome of that collision. "Mobile services" have an impact on individual lives and lifestyles – which is the focus of the book. Even more important, mobile communications and services are permeating almost every sector of the global economy. As a consequence, the processes and tools used in commercial and public enterprises and the role of individuals must be changed forever. While we don't know how this will happen, we do know that these changes will be deep, profound, and long lasting. As more of what we do enters the digital realm the pace of change will likely accelerate. For individuals it will affect the basic essence of their lives – from how they experience entertainment, to how they interact with their friends and family, to how they access education and knowledge, their finances, health care, transportation, and other basic services. For enterprises it will likely be the next revolution in productivity. This book is a good starting point to understand what ingredients are needed and to understand how the revolution might happen.

In closing, I would like to recognize the authors of the book for the contribution they are making. Shoshi Loeb, Thimios Panagos, and Ben Falchuk are researchers in Applied Research at Telcordia. Their charter is to invent the future. Despite the pressures of their full time engagements, they have shown great discipline and the commitment of much of their free time over the past year to shape their creative contributions into this book. Each of them is a creative person, with a distinct style, sense of perspective, and strongly held ideas. They are excellent at what they do. They bring a wealth of experience, insight, and curiosity to this endeavor and they deserve kudos for what they have written.

Adam Drobot

President Advanced Technology Solutions,
CTO, and Member of the Technical Staff,
Telcordia Technologies Inc.

1

INTRODUCTION

"Technology and society shape and reshape each other ceaselessly like tireless meshing gears in a perpetual motion machine."

—Regis McKenna *Real Time: Preparing for the Age of the Never Satisfied Customer.* Harvard Business School Press, 1997

Two of the major attributes of freedom are personal communication and personal mobility. Mobile communication devices have the potential of offering both. Furthermore, mobile devices that compute and communicate have the potential to enable transformative services, for a wide range of industries, that impact personal, professional, communal, cultural, and societal life situations. In order to realize this potential in its fullest extent a large number of new technologies, infrastructure services, business models, social processes, and individual and group behaviors are emerging. Therefore, what we see happening in the mobile services space today is just the beginning. Before we look at the question of where this is heading, let us first examine the technical, business, and social dynamics of the mobile services ecosystem.

1.1 INTRODUCTION

It has long been observed that technology shapes the metaphors we use to describe ourselves, our view of life, our motivation, and our needs and wants. The question

The Fabric of Mobile Services: Software Paradigms and Business Demands, by Shoshana Loeb, Benjamin Falchuk, and Euthimios Panagos

of whether *technology shapes our lives* or whether *we use technology* to more vigorously pursue our characteristic ways of life has been of concern to scholars, spiritual leaders, business people, moralists, and many others for as long as technology in its many forms has been part of our lives. If we focus on the last three centuries and the process of societal modernization, we witness the rise of the corporation, the increase of affluence, the breakdown of the extended family, and the growth of individualism. Regardless of whether these trends shape us or we shape them, they indicate a convergence towards self-absorption and personal freedom.[1] The mobile revolution can be viewed as the next manifestation of this convergence.

The mobile revolution has been enabled by a swirl of new disruptive phenomena, including "experience economy," "social media," "network effect," "long tail," the "big switch," "mashup," "ambient intelligence," "swarming," and "always on". The day-to-day infrastructure of the old order is transforming and giving rise to a new ecosystem and value landscape of innovative businesses. Engineers, software developers, network planners, service developers, content creators, media providers, industrial designers, and a host of others provide the building blocks at the foundation. The ecosystem is dynamic – some of the new businesses succeed in it and some fail.

Where do successful disruptive businesses come from? Some, especially in the venture capital and business communities, believe that you can take any group of smart people and lock them in a room, and they will come up with interesting mobile services ideas. This is only part of what it takes. Even if we assumed for a moment that investors and large ecosystem players had found a way to streamline the creation of new service concepts, the question "What will it take to build a sustainable business around these services?" still remains unanswered. In other words, and at a very high level:

1. Are people going to use the service even if offered at no cost, given the many hundreds of other new services that are becoming available all the time?
2. How does the service fit in the technology value landscape – what pre-existing enablers can support it or add value to it, and at what cost? Such enablers might include location servers, authentication systems, and information services linked into an information "mashup."
3. How is the company offering the service going to sustain itself and, furthermore, attain an expected set of financial goals?

In today's "long tail" of the "markets of one" [AND1], how do we assess the value of existing or new technologies? In strict economic terms, the value society puts upon a piece of technology or a service is determined by its functions. As we are finding out, and will discuss in later chapters in this book, mobile services are mostly an exception to the economic value rule, as functional capabilities are not the only thing that users pay attention to (e.g., [AND], [GOL]). With the advent of the "Experience Economy," where the technology or the services are commoditized and undifferentiated, users find

[1] See [STA], [FIS], and [CAR1] for a perspective.

value in the experience of using the technology and getting the service offer. As Pine and Gilmore put in their breakthrough book *The Experience Economy: Work Is Theater and Every Business a Stage* [PIN]: "Experiences have been always around, but consumers, businesses, and economists lump them into the service sector along with such uneventful activities as dry cleaning, auto repair, wholesale distribution and telephone access. When a person buys a service, he purchases a set of intangible activities carried out on his behalf. But when he buys an experience, he pays to spend time enjoying a series of memorable events that a company stages – as in a theatrical play – to engage him in a personal way" (see also Naisbitt seminal work [NAI]). This emphasis on paying for the intangible experience rather than the tangible goods and services is partly what leads to what economists call irrational behavior [ARI]. This branch of economics is now termed "behavioral economics," and it seems to be governing much of what is happening in the mobile services space and, hence, makes it hard to predict which technologies will survive and which will die soon after being introduced. There are lots of ways to get it wrong; but certain things we do know. For example, mobile services need to be easy to use, simple to operate, and quick to develop and extend. A delicate balance amongst business models, service features and functionality, and price points is required for achieving a wide user adoption and leading to a sustainable business. But in addition to the dynamics of the individual services, we need to consider the interaction of the service with its ecosystem.

In the book *Design Rules: The Power of Modularity* [BAL], the authors discuss the concept of a "value landscape" as it was first introduced by Alchian in his seminal paper on "Uncertainty, Evolution and Economic Theory" [ALC]. The economic system is viewed as a very large value landscape that rewards the creation of new technologies, products, services, new businesses, and, ultimately, new knowledge. Within this landscape, value is not only "the height" or "latitude" of a specific technology or service but also a force that pulls both technologists and companies "upward" in the landscape. This perception of value is what compels entrepreneurs and investors to create companies that seek to both improve existing technologies and create new ones. If the success of a product or a service were predictable, the "force" of market value would pull the ecosystem players unambiguously in one direction. The process of new product or service creation would then be a straightforward deductive exercise focused on implementation. But as we have witnessed in the high-tech market in general, and during the dot-com boom in particular, location of the "high peak" in the value landscape is uncertain, and when revealed, even surprising. Therefore, the steps that need taking and the time needed to reach a given peak are unpredictable as well. In some cases, even the "direction" of the nearest peak may be unknown. The fundamental uncertainty, combined with the inherent complexity of mobile technology, products and services (the high dimensionality of the space), makes certain approaches to planning and implementing a new offering in the mobile services space much more effective than others. Specifically, in the presence of uncertainty and complexity, modular design becomes very valuable. Modularity enables design options that can radically change the market value of a given technology. This is why, as we will see later in the book, the approach taken across the board in the mobile services marketplace is to combine enablers, features, and functionality to create new offerings that leverage modularity and re-use at all levels. This includes, for example, the "mashup" paradigm in which a new application or service is created through the "gluing together" of existing information and services running on remote servers.

The evolution of mobile services and the value landscape they create entail the co-evolution of several industries that did not have much prior opportunity to converge. Today, we are witnessing the collision of the mobile device industry, the media and content industry (e.g., Web 2.0), the computing industry, the software applications industry, and the consumer electronics industry, as well as the automotive, healthcare, and many other vertical specific industries. With a collision of this magnitude, there is no telling what the result is going to be. For instance, are mobile phones with cameras on a convergence path with Wi-Fi enabled cameras? What about telephony-enabled laptops and smart phones with sufficient computing and processing power?

The purpose of this book is to present mobile services from the perspective of the unique "fabric" of its ecosystem and its essential enablers. It is the enablers that we focus on not necessarily the services themselves, although we mention services as well as needed. We look at a few representative elements of the fabric, grouped into a few key topics:

1. Simplicity and user experience;
2. The "always on" challenge;
3. Opportunistic mobile service;
4. Design patterns.

In the rest of this chapter, we first briefly discuss what we mean by next generation mobile services, discuss three of the tangible components of the mobile services value landscape, the mobile device, the mobile Web, and the communication network, and then look at a service example. We then discuss the "social rules of mobile services" and overview the contents of this book.

1.2 UBIQUITOUS COMPUTING AND NEXT GENERATION MOBILE SERVICES

The mobile device today is an integral part of our lives. Some studies indicate that people who leave home accidently without the device usually discover this fact within a few short hours, much quicker than people realize their wallet is missing! Bloggers musing, almost nostalgically, about the days without the mobile phone (or the Web) are now commonplace. They wonder: "how did we manage to do the shopping without the option to call home to say that the store is out of bananas and what should we get instead?" or "What did teens do without being able to share every moment with someone remote and mobile?"

As is the case with many concepts that finally hit the "mass market," what we are seeing emerging today in the mobile services market was envisioned long ago, as part of the vision of ubiquitous computing. Ubiquitous computing [SCH] (also see recent reviews in [GRE] and [MOR]) portrays a world in which computing power is so distributed throughout the environment that it becomes an integral, invisible part of life. The highway, the car, the living room, and the garment become sites of information monitoring, processing, mediation, and transmission. Simplicity dominates the user experience: interactions feel natural and spontaneous. The technology "smoothes the edges of modern life," guides us, intervenes, and gently reminds us. In this paradigm, all the information that today sits in isolation on our phones, personal computers, and Web sites becomes accessible from just about everywhere, at any time, and is exploited to deliver value in a manner appropriate to our current location and context.

That is the vision, the end game, the objective that the mobile services industry is striving for explicitly and implicitly. To make this vision a reality, we need further innovation in several areas that touch the mobile devices, the supporting networks, the application infrastructures, and service enablers. This includes, but is not limited to, the following:

- Smart mobile devices, some with nontraditional form factors (e.g., wearable and organic devices);
- Wireless sensors embedded in our environment and artifacts;
- Seamless communication and computing infrastructure supporting interoperability between wired and wireless networks;
- Service-oriented architecture for flexible and dynamic service orchestration;
- Semantic Web for machine readable Web, and application enablers such as presence and location;
- Dynamic and massively distributed programmable device networks;
- User-centric functionality including artificial intelligence and multimodal interaction (e.g., speech recognition and synthesis);
- Awareness of the user context with all of its complexities and nuances;
- Ways to ensure privacy and personal security.

More than the technical capabilities alone, ubiquitous computing in its current incarnation, albeit deficient, creates an experience users value. "It is no longer about technology," users say, "it is about the experience of being always connected." In the economic value landscape, experience is an intangible parameter and cannot be measured directly using traditional economic tools. How do you put a price tag on "smoothing the edges of life?" and, more importantly, which technologies do you use to make this happen?

1.3 THE MOBILE WEB

The promise of the mobile Web is to provide rich and personalized user experience that goes beyond what has been available so far using fixed personal computers and mobile devices. However, so far, the mobile Web has largely fallen short of expectations. It

Table 1-1: Web 1.0 ⟶ Web 2.0

Web 1.0 Entity or Trend	Web 2.0 Equivalent
DoubleClick	Google AdSense
Britanica Online	Wikipedia
Personal Websites	Blogging
Screen scraping	Web services
Publishing	Participation
Content management systems	Wikis
Directories (taxonomy)	Tagging ("folksonomy")
Stickiness	Syndication

Source: Adapted from [ORE]

has been noted that the current mobile Web is an abbreviated version of Web 1.0, and a large fraction of mobile Web users are currently denied the features of Web 2.0.

How did Web 2.0[2] come about? The concept of Web 2.0 began with a conference brainstorming session between O'Reilly and MediaLive International around 2004. It was noted that far from having "crashed," the Web was more important than ever, with exciting new applications and sites popping up with surprising regularity. Some of the companies that had survived the collapse seemed to have some things in common. Could it be that the dot-com collapse marked some kind of turning point for the Web? The consensus was that it did, and so Web 2.0 was born. Table 1-1 highlights some of the key differences between prototypical Web 1.0 and Web 2.0 sites and concepts.

The early days of the dot-com revolution produced two seemingly conflicting activities. On one hand, it was clear that "Webifying" almost any business model could lead to better value for the end user and greater efficiencies for the enterprise. At the same time, there was a surge of creativity when it came to using the Web to transform businesses and lives. Some of the service concepts were slow to take off due to technology gaps and inadequate business models, or inappropriate supply chain management and processes. For example, social networking sites – while useful – were not nearly as compelling before there was a large amount of ancillary content – such as blog entries, home videos, and original music – that could be linked to and from users. A similar mindset can be found today around "mobilizing" services in the areas of telecommunication, media and entertainment, advertising, gaming, healthcare, and automotive telematics, as well as enterprise services such as work force management, sales force management, customer relationship management, and supply chain management. Recent service offerings by wireless carriers[3] as well as new browser technologies[4] are attempts to remedy the situation. Technical and usability problems still exist but are likely to be resolved in the

[2]Source: http://www.oreillynet.com/pub/a/oreilly/tim/news/2005/09/30/what-is-Web-20.html

[3]AT&T and Verizon Wireless provide access to multiple social networking sites using Intercasting Corp's technology.

[4]Opera Mini, Skyfire, iPhone Safari.

Table 1-2: Mobile data users in the US (2008)

Wireless Operator	Wireless Technology	Wireless Subscribers	% Data Subscribers
AT&T	GSM	71.4M	28% (20.0M)
T-Mobile	GSM	30.8M	25% (07.7M)
Verizon Wireless	CDMA	67.2M	31% (20.8M)
Sprint	CDMA	52.8M	25% (13.2M)

future. What may not be resolved that soon is the introduction of new compelling mobile Web services that will attract and sustain mass-market users beyond what is being offered today. One of the issues with mobile Web applications is achieving a critical mass to both benefit from the social networking "network effect" and provide a solid business model for the service providers. One of the gating factors is how many people today actually pay for data services on their mobile phone, as data services are needed in order to access the mobile web. Table 1-2 shows the current number of mobile services users in the United States and what percentage of them have data services that are required in order to access the Web from the mobile device as well as to receive email. The percentage of these users is only 25%, which means that there are great opportunities for growth ahead. Is this sufficient for a critical mass?

All this being said, a 2008 Nielsen Mobile Company[5] report says that mobile Web users have reached a critical mass:

> "The US is the most tech savvy nation with nearly 40 million Americans – 16% of all US mobile users – using their handset to browse on the move. In the US, the number of people using the mobile Web has increased from 22.4 million in 2006 to more than 40 million today. However, roughly 95 million Americans were paying for mobile Web access but did not necessarily use the service as mobile internet is often included as part of a larger mobile media package. Users may be either unaware or disinterested in the internet access that is provided. This is not true of owners of certain handsets, such as the first generation iPhone. 82% of iPhone owners access the mobile internet, making them five times as likely to do so as the average mobile consumer. The second-generation iPhone comes with 3G, allowing faster access to the Web... However, the firm found that browsing habits differed between a PC and the small screen. PC internet users visit more than 100 domains per month, on average. By contrast, the average mobile internet user in the US visited 6.4 individual Websites per month. UK use was slightly less at 5.5 per month, whilst Italian users visit 8.2 per month on average. This is due to Italians using handsets that are more sophisticated. Growth like this means the mobile Web is now a viable option for big business."

The more industry, technology, and market reports one reads the foggier the picture of what exactly is going on in the mobile market today may get, and the picture changes daily. In Chapter 2, we go into the details of why the market appears so uncertain and what the forces are (some of them opposing forces) that shape its somewhat unstable dynamics. For the purposes of our discussion here, we will take a brief high-level look at some of these dynamics.

[5]http://www.nielsenmobile.com/

1.4 EMERGING MOBILE SERVICES ENVIRONMENTS – NETWORK VIEW

To offer a mobile service today, you have to be of one of two minds. Either you think "Web and Internet," or you think "telcos and cellular." If you think Web, you tend not to worry too much about dealings with telcos, as you believe that your services will be offered to the mobile device straight over the Web, completely bypassing any telco control. In reality, things may not always be as simple, at least not for now. In the following, we will take a quick look at the two approaches to offering services.

The telco view on offering a next generation services environment is based on the IP Multimedia Sub-system (IMS) [CAM]. IMS as defined by 3GPP was originally architected as a "horizontal" (i.e., cross-cutting, or spanning) service control layer for rapid deployment of IP multimedia services. However, given market conditions, which favor a modular service enablement environment rather than an all-encompassing and, perhaps, constraining services layer, IMS is being rolled out as a middleware services infrastructure supporting several services enablers and reusable service enablement functionality, including the following:

- Presence and location for mobile devices;
- User and device identity and preferences;
- Service plan information.

In principle, these enablers allow service creators to build differentiated, feature-rich IP voice and multimedia mobile services, such as:

- Presence and location enabled voice and integrated messaging services with mobility management and social networking interaction capabilities;
- Adaptive device-specific video and multimedia services;
- Trusted third-party e-commerce payment services with micro billing and bill back to users' subscriptions plans.

Figure 1-1 illustrates a layered view of IMS, with each layer representing a separate business and technical concern. The key value proposition of IMS is the decoupling of the services from the access network layers, thereby simplifying considerably the process of creation and introduction of new services. Web Services, a "plumbing" technology defined by W3C, are a big part of another related middleware called Parlay X Web Services, which expose network capabilities to developers as Web Services, thereby making it easier for operators and developers to leverage the network.

Table 1-3 illustrates some of the differences between the Web and the telco-oriented services methodologies. As these methodologies were developed with different technical and business goals in mind, it is not surprising that they represent opposing points on many dimensions. Telco architecture depends upon "event-oriented" network signaling protocols (e.g., SS7, SIP), making it ideal for the creation,

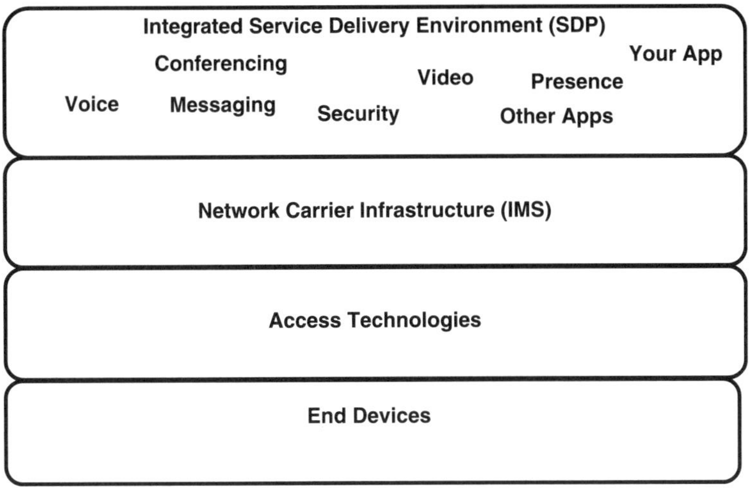

Figure 1-1: A simplified layered view of the converged mobile service ecosystem. What matters to service creators are the top (applications), which may use enablers from the second, and bottom (devices) layers.

control, and delivery of real-time network services such as voice/video-over-IP, push-to-talk, seamless voice mobility services, etc. A service-oriented architecture (SOA), on the other hand, is based upon XML and Web Services frameworks, which makes it ideally suited for longer latency and feature-rich "transaction-oriented" service creation and delivery. Examples of transaction-oriented services enabled by SOA could include: 1) ad-supported mobile multimedia streaming and download services, with revenue sharing among the content providers, operators, and handset makers, and 2) comparative on-line shopping services with micro payments and location-enabled social networking.

To get the best of both worlds, IMS operators are now utilizing SOA-based Services Delivery Platforms (SDP) implemented over the IMS layer (see Fig. 1-1)

Table 1-3: The differences between Web services and traditional telco services

	Web Services	Traditional Telco Service
Architecture	Transaction-based	Event-based
Latency	Medium to high	Low
Transaction time	Long	Short
Complexity	Low	High
Programming	Scripting	Coding
Enablers	E-commerce, user metadata, ad-insertion, gaming, media sharing, mapping, dynamic charging, LBS, personalization	Routing, screening, authentication, charging, messaging, presence, real-time communications

Source: Adapted from [MOD]

to help streamline "service creation" and, in turn, the introduction of new IP services. They are also using IMS standards for underlying service enabling and delivery aspects, such as charging, device presence, location, subscriber databases, and hierarchical levels of network and services security. The dynamic aspect of new service creation and delivery is especially important to the operator, given the highly competitive IP service provider environment, which necessitates rapid new service creation and deployment cycles. These cycles are increasingly hinged upon "access independent" approaches, short (12–18 month) payback periods, and low marginal costs of new service introduction. This may allow the network operators to leverage their SDP/Web services platforms to engage in profitable value chain partnerships with Web companies and content/e-commerce providers, involving revenue sharing and transactions fees. The partnerships may involve offering the value chain partners real-time "user information," such as user identity, security credentials, handset presence, service/feature characteristics, location, and demographics. They also may enable the introduction of targeted advertising and transaction-based business models for influencing users' demand and adoption of premium 3G services. To summarize, the IMS/SDP approach may provide wireless operators with the following business benefits:

- *Support Services Enablers* – It is reasonable to expect that there will be no single killer application but a wide variety of interesting IP-based services whose features can be aggregated to form even newer services. For example, combining a presence and location-based voice service with buddy lists and media may result in a new social networking application. This approach ensures that overall revenues are not dependent on the most popular services – which change often – but are distributed among a wide range of services.
- *Experiment with New Business Models* – The SDP/IMS platform is designed to enable operators to implement new services delivery business models, especially ad-supported and transactions-based models. The new business models hold promise for operators looking to 1) introduce feature-rich wireless services, e.g., presence and location-enabled voice and multimedia services, supported by targeted ad insertion based on the users' identities and demographics; and 2) generate transaction fees from ecommerce activities that are enabled via their services. Using an experimental approach towards new services definition and trialing is clearly a trend that will lead to rapidly discover the next generation of sticky and profitable services.
- *Create an "Open" Environment for Ecosystem Providers* – Most network operators have recognized that they alone cannot create and deliver *every* valuable service. An SDP/IMS infrastructure enables an "open" applications services creation and deployment environment that is conducive to incubating an open-ended ecosystem (value chain) comprising third-party content providers, service providers, applications developers, and device and systems vendors. Note that, in this case, the network operator is still in the picture, but as we will discuss in Chapter 2, they control less of the ecosystem.

1.5 THE MOBILE DEVICE ENVIRONMENT

Mobile devices today have progressed remarkably from mere voice communication machines (i.e., telephones) into state-of-the-art platforms that can send and receive e-mails, take mega-pixel digital pictures, access Websites, monitor biometric information, and so on. Furthermore, modern mobile devices allow service creation through such operating systems and platforms as Symbian, Linux, Java ME, Android, Widgets, and iPhone SDKs. It is expected that these platforms and their future versions will steadily gain momentum and support from a worldwide ecosystem of application developers, device vendors, and wireless network operators.

Steps toward weakening the hold that the phone companies have over the mobile device applications continue to be taken. In the year 2000, the Java profiles (e.g., MIDP) implemented on Connected Limited Device Configuration (CLDC) opened up the applications playing field to Java developers. More recently, the formation of the Open Handset Alliance Project (which includes mobile operators, handset manufacturers, software companies, and semiconductor companies, and as members)[6] is doing a similar thing. All the tools necessary for the development of applications for the platform are free and open, and the devices will treat (almost) all applications equally. This will create a disruption in the value landscape and Google, which is the driving force behind the effort, will benefit from the wider availability of advanced applications on mobile devices.

Due to the richness of technology, business context, and user context, mobile devices are viewed from a different perspective by different industry plays. Below are some examples of these perspectives:

1. The telecommunications industry traditionally views the mobile device as a small phone you can take with you.
2. The media industry views the mobile device as the *fourth screen* on which media content can be viewed. It comes after the movie theatre, the television, and the personal computer.
3. The computer industry views the mobile device as an extension to the personal computer.
4. The gaming industry views the mobile device as an extension to the gaming console.
5. The healthcare and well being industries view the mobile device as an extension to the doctor's office, the hospital, the nurse, or the counselor. It can support reminding services for medical compliance to a glucose monitor to a personal well-being coach.
6. The automotive telematics industry views the mobile device as an extension of the car and use it as a gateway between the car and a wide range of connected services.
7. Many other verticals, such as real estate, are coming up with unique applications that add new dimensions to their business.

[6]See: www.mobilehandsetalliance.com

Looking ahead to the next one or two years one can expect device features such as large high-resolution screens, powerful processors, larger storage space, considerable battery life, intuitive input stream, video phone, compass, USB port, accelerometers, and software-defined radio. These features will enable gestural input and innovative applications and many other features.

1.6 THE PERSONAL ASSISTANT EXAMPLE

One of the hardest applications to put together is one of the first ones that comes to mind when one thinks of an advanced futuristic technology – the *personal assistant*, the all-intelligent system that can know what you want even before you do. In this section, we devise a thought experiment in which we construct such an application, and then add more complex functionality to the basic service it provides. This thought-experiment is aimed at highlighting enablers and architectural options.

The personal assistant application is intended to find information for the user when it is most needed (e.g., a plumber to come for some fix-up, a restaurant on our way). In the simplest case, the system that implements this application is nothing but a straightforward directory lookup – looking up the service provider we need in a directory. The directory may be populated manually through an administrative interface and may be accessed through a lookup function. A user interface is required to input requests and return answers. The interface is then enhanced with some scheduling and calendaring functions, allowing the user to specify the time he needs the service and to schedule recurring services. It may also allow the user to specify the location of the service and other attributes such as cost.

As the next step up in sophistication, the system manages the user's profile, as well as richer descriptions of service provider capabilities. At this stage, the system is still doing simple matching between user profiles and the service providers' directory; but will need an additional interface to input and manage the profile. Next, the system will incorporate user feedback on past selections. The feedback requires an additional user interface to give the feedback and another to consult it. Next, the system hooks up to other information systems, allowing it to understand provider rankings.

Next, the system can apply more sophisticated selection criteria and use collaborative filtering techniques in order to guess, or predict, which services the user might like based on what other people with similar preferences like. This feature may also help increase the application's up-take in a community.

After introducing the notion of a community of end-users, the system now can incorporate the notion of a community of service providers and allow them to introduce their own profiles and preferences for matching in a more dynamic way than a static "yellow page-like" style. Sophisticated recommendation and notification algorithms will be needed to meet the requirements described above. Operationally, once a service request has been initiated, the system parses it and matches it with information from its database to identify qualified service provider candidates.

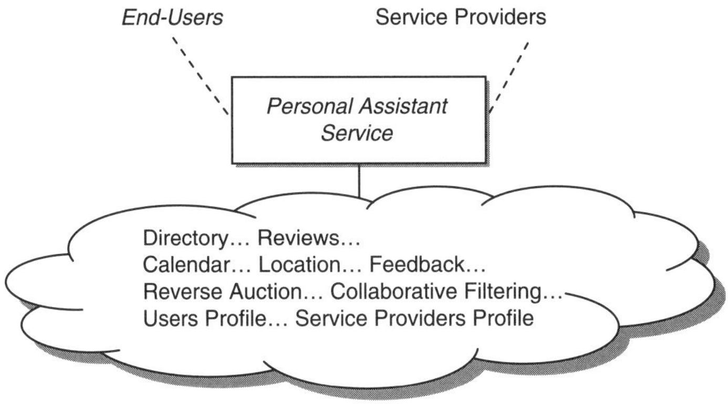

Figure 1-2: Personalized assistant example.

Recommendation algorithms will further filter and sort the results of the service request so that the list of results can be presented in a manner that is meaningful and attractive to use. After a selection is made, the system notifies the chosen service provider using one or more methods of communications (e.g., e-mail, Web-based, SMS, traditional telephony).

Figure 1-2 illustrates the system at a high level. Now things get a little more interesting as context comes into play. Imagine that the user carries a mobile device and allows the application to know his whereabouts and imagine further that the application knows personal information such as his calendar and to-do lists and that he is now "driving from Point A to Point B". This version of the application becomes complex and "context aware" [SCH].

An interesting next step would be to evolve the system from a Personal Assistant into a Market Maker (e.g., [NIS]). One way this can happen is to add an auctioning function to the infrastructure in which the service providers might bid on the right to satisfy a user's service request, perhaps through a reverse auction paradigm (as implemented by a "market maker"). An important thing to notice here is that advertisers can play the role of service providers. Here is how the different parties benefit from the service:

- Market makers may benefit by charging a fee to service providers to participate in the auction and, in some cases, increase their revenue due to larger profit margins;
- Local service providers benefit from participating in the program by gaining access to revenue generating opportunities that they may not be exposed to otherwise. Furthermore, service providers who win the auction process may increase revenues through follow-on services.

In the rest of the book, we will be revisiting in much more detail aspects of this example.

1.7 THE SOCIAL RULES OF MOBILE SERVICES

So far in this chapter, we have seen selective glimpses of the dynamic technology and business value landscape. The pace of change in the market will soon make obsolete much of what is being written today. Therefore, in this book we attempt to find some insights that we believe are universal about the technology, the business, and our lives; insights that may guide our understanding of service evolution. Our cognitive abilities are no different from what they were over thousands of years ago (with some believing that they have deteriorated due to technology [CAR]). Furthermore, if we check the recorded history we will find out that our complaints about life and the difficulties of life have not changed much. Therefore, as we indicated previously, the human part is *the* gating factor in our acceptance of anything new, including new mobile services. Consequently, perhaps there are some insights hidden in the swirl of "the mobile services market." Let's examine some of these insights. A good starting point is to look at Rachel Hinman's Adaptive Path Blog, [HIN], and examine some examples of how mobile devices are changing the social rules:

1. It is acceptable to talk to yourself in public places when using a Bluetooth headset or speaker feature on the mobile phone. This, of course, sometimes creates a backlash.[7]
2. Individuals show addictive "Crackberry behavior" that may be alarming to friends and family in spite of the enormous increase in personal and business productivity associated with the "always on" device.
3. One is always perceived to be available now that one has a mobile phone, and not responding sometimes creates negative connotations.

This last point is an example of technology creating a situation (e.g., being always connected) that then leads to a problem when one is not immediately responsive. There might be many technical solutions to solve this, but each may in turn create a fresh set of new problems (privacy-related and so on).

According to Hinman, social rules are based on our understanding of expectations – expectations about ourselves and about how we want people to engage with us. In the early days of any major disruptive technology, social misalignments and issues arise (see also [FIS], [STA]). It is clear that we associate different expectations to the various communication channels. When we call someone, we expect them to answer and if they do not answer, we leave a voicemail and expect them to get back to us. The rules for e-mail and text messaging are slightly different – often the social rules are personal and reflective of the relationship the sender and receiver have – or the social contract they share. When someone does not answer the phone, there is a set of likely explanations: the receiver is busy, the user is not within reach of their phone. A conflict seems to occur when expectations are not in alignment. For example, technology such as GPS positioning – which we will

[7]See Fred Wilson's "the no voice internet" http://www.avc.com/a_vc/2008/09/the-no-voice-in.html

cover later on in the book – or an Instant Messaging status message of "busy" adds a layer of complexity. Using these tools, a sender can know more about the receiver's state and adjust their expectations accordingly perhaps at the expense of the receiver's privacy. For a service to be widely adopted, the service designer and operator must be sensitive to all these social rules. For example, as we see later, even if the service operator may know the full context of the user, it may not want to reveal it all. The service must:

- Allow people to do what they need to do to retain social contracts and privacy;
- Support different levels of profiles and different levels of engagement to reflect the variety of expectations people have for various relationships.

Hinman and her workshop participants state that mobility implies changing contexts and changing interruptability. When designing a mobile service, assessing user interruptability may be key to user acceptance and it all comes down to the question: *Who is responsible for managing the user attention and intention?* Is it the user through explicit protective measures expressed via preference lists, calendar entries, to-do lists, profiles, or application configuration files, or is it the responsibility of the service designers or the service operators? As an example, when people first signed up to Twitter,[8] which allows them to report to their friends what they are doing at any moment, they chose to follow lots of friends, but they eventually turned off the service because they were inundated with too many "tweets." As Hinman points out, was the problem a flaw in the design of Twitter or was the problem in that people did not fundamentally understand the "rules" of social networks on mobile? This is one of the hardest questions facing the market today. Fundamentally, Hinman argues users should have no barrier to what they can do with a service, and that ultimately the burden for providing controls should be embedded in the system. However, as Hinman indicates it is often difficult to predict what will happen and design appropriately. Therefore, in the end one is not really offering services to users, but offering services to users who are embedded in a context. The issue is recognizing the different roles that people have and the communication channel they use – SMS, mobile Web, voice, applications. Depending on the context of your application, expectations are different. Measuring intangible user experience, like measuring any intangible (see for example methodologies discussed in [HUB]), is still a challenge.

The problem of capturing user context from various available information sources such as calendar entries, to-do lists, e-mail interactions, and phone calls is hard (e.g., [WAT]). Progress has been made on this problem, usually under the umbrella of intelligent digital personal assistance as we discussed in the previous section. Software agents, when implemented effectively, help users with everyday duties like managing timetables, organizing travel, and so on. Mobile intelligent agents may soon enable applications likely to create a lot of interest, both among operators and users, in fields as diverse as interactive personalized content, e-commerce, or gaming.

[8] www.twitter.com

1.8 THE FOCUS OF THIS BOOK

In this book, we attempt to blend several perspectives on the evolution of the mobile application technology, services, and, ultimately, markets. This is not an easy task as we are bound to occasionally tilt too much towards one perspective. It is our hope, however, that the temporary unevenness will smooth out through the flow of the book as a whole.

In Chapter 2, we first paint with large brush strokes the picture of the mobile services market as a highly complex system that is in a state of transition. This reflects at the macro level with several large industries such as telecommunication, computing, and entertainment on a collision/convergence path, each with its own momentum. At the same time, at the micro level, the level of the individual user, the system exhibits what economists call irrational behavior, where "user experience," which is an intangible, counts more than tangibles. We then proceed to discuss the notion of a value chain as an indication of a business model and show how the value chain warps as the market evolves, when players either expand their offerings into neighboring parts of the value chain or buy companies that dominate a market segment. This is followed by a discussion of the two worlds that the mobile services emerged out of from the network viewpoint – telecom and Web – and proceed to look at current innovative business models. We conclude with a specific example of personalized mobile service and its various business dilemmas.

Location-based services have several key technical components. In Chapter 3, we devote a significant discussion to the technical enablers that make such services possible. This includes an overview of the cellular, radio, and satellite communications systems with a particular focus on how their respective positioning technologies and accuracy plays a role in mobile services. We then examine the functional components that comprise today's location-based services (LBS) deployments. For example, mapping, geo-coding, content filtering, and service logic are critical pieces that operators need to deploy in order to run LBS successfully.

In Chapter 4, we shift perspective and turn to a discussion of the merits and principles of simplicity and user experience in mobile services. This includes a look at what the new expectations are for mobile user experience, how needs of mobile users differ across segments, and how mobile hardware and services are being designed with simplicity in mind (and how they may *not* be). We complete this discussion with real-world examples of mobile simplicity and user experience, focusing on some real products and services.

To meet the needs and expectations of mobile users, mobile services have "always-on" requirements. In Chapter 5, we give this challenge an in-depth look by first examining why it is so important for mobile services to be highly available and perceived as always on. We then look at the infrastructure and IT components and concerns that provide always-on capability and the important distinctions between scaling "up" versus "out."

In Chapter 6, we look at an emerging paradigm for mobile services that many refer to as "opportunism." Although this term often refers to link layer aspects of communications, there is a larger sense for the term in the mobile services realm. We will explain the term, why it is an important new paradigm, and what it takes to support it successfully. Such support technologies include Semantic Web concepts [BER], artificial intelligence, agent-based systems, and others. We will provide detailed examples of opportunistic services that are currently in research or in real deployments; this will give the reader a clear picture of the scope and function of such services. For the benefit of planners, we outline what is still missing (e.g., in standards, in markets, and in technology) and make predictions on where opportunism is heading.

Designing and implementing mobile services takes special insight and skills. Today's current art is a mix of older techniques and software patterns with emerging paradigms and best practices. As mobile services continue to diverge from what services used to look like, newer design patterns will dominate. In Chapter 7, we will take an in-depth look at the old and new patterns that form today's mobile services view. These will include architectural patterns such as Model-View-Controller and client-server, as well as issues relating to User Interface structure and design (e.g., wizard, slideshow, and pagination techniques). Data distribution and resource management will be presented as key issues. Widely seen as a fundamental enabler of future mobile services, event-driven architectures will be also examined.

To wind up the book, Chapter 8 examines future trends, advanced research, and cool services. The players, the challenges, and the current state-of-the-art will be studied. In that vein, we survey some selected advanced new research in mobile services that will give readers a good idea of the sort of innovation that is around the corner.

2

THE BUSINESS OF MOBILE SERVICES

"... there is a huge difference between cheap and free. Give a product away and it can go viral. Charge a single cent for it and you're in an entirely different business, one of clawing and scratching for every customer ..."

—Chris Anderson, "Free! Why $0.00 Is the Future of Business."
Wired, Feb. 2008.

Business analysts are burning the midnight oil trying to figure out where the multi-faceted business of mobile services is headed and who will come out the winner in the many technology and business frontlines. Technology investors and ecosystem players need these insights in order to make decisions; decisions that may make or break their business two to five years down the road: What will be the coverage of the 3G network? How many smartphones will there be? What will the killer app be and will one be able to control or profit from it? What role will the mobile network operators play? Will the Android platform transform the market even more than iPhone? Which operating system will dominate the mobile device market?

As we discussed in Chapter 1, technological innovation and human behavior are in a never-ending dance. Technological innovation disrupts behavior and that, in turn, creates business opportunities and openings for the next wave of innovation. This is not new and is not going to end any time soon. As Freud noted in his seminal work, *Civilization and Its Discontents*, we create and use new technologies that solve problems that previous technologies created, which then create the opportunity for the next level of innovative "fixes" [FRE]. The example he used was that of the

The Fabric of Mobile Services: Software Paradigms and Business Demands, by Shoshana Loeb, Benjamin Falchuk, and Euthimios Panagos

telegraph. In his view, the telegraph was solving a problem that was created by the train, which took people away from each other – no train, no need for the telegraph. According to Freud, our drive to innovate originates from a fundamental need that can never be satisfied, and so, as a society, we are stuck with a perpetual fixing cycle. But even if we grasp that there is an underlying never-ending drive to innovate, we are still left with the question of the way new technologies become known and reach mass market appeal because, ultimately, for a technology to have an impact, it must be adopted. The value chain ends with the customer.

What affects customers? How do customers decide what to adopt and when? In his transformative national bestseller *The Tipping Point: How Little Things Can Make a Big Difference*, Malcolm Gladwell popularizes earlier scientific work in epidemiology, psychology, sociology, and group dynamics and applies these insights to an array of social behaviors and cultural trends [GLA]. The social behaviors and cultural trends include word-of-mouth spread of products and services that are critical to the success of mobile services and their technological underpinnings. The key here is the book's subtitle, "How little things can make a big difference," and these little things more often than not determine the success of a product and the direction of an industry, from nuances of context in which the service is introduced to the influence of key people in a social circle.

If, here as authors, we attempted to take a detailed snapshot of the mobile services market today, we would have produced a laundry list of technologies, business events, and speculations that may end up being inconsistent and incoherent, not to mention obsolete. If you browse the daily technology and business blogs, you can see what we mean – it is hard to make sense of all that is happening and sort through what appear to be opposing actions, opinions, and customer realities. We believe that taking this approach would not have served our readers and, furthermore, the information we would have been providing, including the various industry projections, might have become obsolete or wrong even before the first copy of this book saw the press. Bob Lucky captured this sentiment best in his recent column in *IEEE Spectrum*, entitled "The Elusive Future" [LUC]. He was looking at whether we can predict what is coming next in technology. Here is an excerpt:

> "In countless panel discussions on the future of technology, I'm not sure I ever got anything right. As I look back on technological progress, I experience first retrospective surprise, then surprise that I'm surprised, because it all crept up on me when I wasn't looking. How can something like Google feel so inevitable and yet be impossible to predict?" [LUC]

What was it about Google that made it so impossible to predict? One perspective on Google's disruptive effect is to see it as a "comparable" to Netscape. Just like Netscape was the big business disruption at the foundation of Web 1.0, so was Google for Web 2.0. As Tim O'Reilly points out [ORE], Netscape viewed the "Web as a platform" with its flagship product – the browser. The following summarizes Netscape's business approach:

- Promoted the "Webtop" to replace the "desktop" and viewed itself as extending Microsoft's desktop paradigm into the Web, similar to when the automobile was framed as a horseless carriage;

- Looked for control over standards for displaying content and applications on the browser;
- Planned to make money on high-priced application server products that information/application providers would buy for pushing content to the browser.

In the end, both the Web browsers and the Web servers became commodities and value moved up the chain to services delivered over the Web platform. By contrast, Google started as a pure-play Web application using the Software as a Service (SaaS) business model. Google:

- Never sold, licensed, or packaged software, but delivered a service based on their software, with customers paying directly or indirectly for the use of the service;
- Has no external scheduled software release (with the exception of Android), just continuous (often daily) improvements, upgrades, and beta releases;
- Never had to deal with costly porting to different computing platforms. Rather, Google developed a massively scalable collection of commodity PCs running Open Source OS and homegrown apps and utilities.

Therefore, the "sneaking up" on us that Google did was a complex process that involved various paradigm shifts including a change in a business model and technology infrastructure and in the way the "product" is being delivered to the user. In his book *The Search: How Google and Its Rivals Rewrote the Rules of Business and Transformed Our Culture*, Battelle [BAT] states that: "Google garnered impressive word of mouth among its users for one reason: it worked. Not only did its PageRank-based algorithms produce delightfully relevant results, but they did it with impressive speed, and the service never showed signs of buckling under the exponential growth it was experiencing."

How many such Google-like disruptions are in store for us in the area of mobile services? One can be certain, given the observed dynamics of innovation and disruption, that there are further disruptive companies on the horizon. However, identifying such companies and their products, and estimating when they are going to sneak up and disrupt our markets, is impossible to predict. This is true not only because of the high level of uncertainty in the market, which is definitely contributing to the "fog" and the ability of companies to sneak up, but also, trivially, by definition – we cannot predict a surprise! How then can we visualize where we are and where we are going in the market?

If we view the mobile services market as a giant "convergence" or "collision" of several very large industries such as telecommunications, computing, entertainment, and the like, how can we "simplify" it conceptually so we can begin to make some sense out of it all? One thought experiment is to imagine the colliding markets as large galaxies coming together, each with its own dynamics and inner structure, each obeying well-defined rules. Although the laws governing the elements of the galaxies are well known, having them on a collision course creates new very complex behaviors. The colliding of galaxies ends up producing new galaxies after a period of apparent chaotic dynamic behavior [LOE]. If one observes only the transient behavior caused by the realignment of the galaxies, the picture may seem surprising and overwhelming. By realizing that something new is emerging out of the old order, we can at least put things in perspective. The same is true when we observe our emerging mobile services market. From the juxtaposition of the customers, applications, and momentum of the different participating industries, very interesting, perhaps temporary phenomena emerge. So, in this chapter, instead of looking incrementally ahead and trying to analyze phenomena that may end up being side effects of a new market that is shaping itself from very large existing markets, we attempt to shed light on some of the major forces at play. Ultimately, it is the balance between these forces that will help the market to stabilize at a new plateau.

2.1 THE ECOSYSTEM AND THE VALUE CHAIN

In their book *The Keystone Advantage* [IAN], Iansiti and Levien state, "The evolution of business ecosystems, their ubiquity, and the reliance on other organizations to accomplish all but the simplest tasks have changed the nature of business. From a world of vertical integration, internal capabilities, internal R&D, internal projects and internal infrastructure, we have come to a world of mutual dependencies, distributed innovation, technology integration, trading collaboration and on-demand capabilities." This statement holds particularly true for the worlds of telecommunications and computing, which are on a collision/convergence course when viewed from the perspective of mobiles services.

A way to observe the ensuing disruptions in the marketplace is to watch how the value chain is being reshaped. Keith Willetts, chairman of the TeleManagement Forum, observes the business environment from the perspective of the Telecom marketplace (which is one of the "galaxies" on the collision course) and says: "The Telecom Industry stands at a crossroads. It's in an expanding market sector with more complexity than anyone could have imagined."[1] He argues that the value chain is no longer linear and simple. For example, in the content area one does not find content providers such as Disney or YouTube handing off content to service providers, who in turn hand it off to home gateway providers. "In reality, there are major changes happening simultaneously that are helping to transform telecom forever and twist the traditional value chain in new ways." In Figure 2-1, we show a generic value chain for mobile service.

Elements in the value chain do not necessarily correspond to business boundaries as the wireless network operator (e.g., Verizon Wireless) can also play the role of service aggregator or application platform provider. The same holds true for other players who repeatedly expand their presence along the value chain (e.g., Apple, Google, and Amazon). They do so by acquiring companies that are active in parts of the value chain

[1] Keith Willetts, "The New Content Value Chain," Telephony On Line, http://telephonyonline.com/software/commentary/telecom_new_content_value/, November 5, 2007.

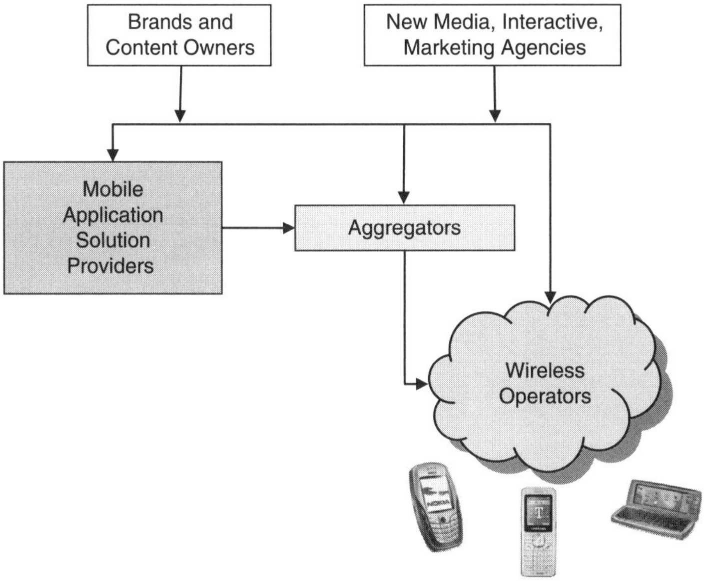

Figure 2-1: An example of a "value chain" for mobile marketing. It highlights the connected nature of the "chain" of content providers, marketing campaign managers, aggregators, and the wireless solutions and delivery channels.

where they need a presence or that enhance their own products and services. Examples are many – Apple iTunes, iPhone and App Store,[2] Google Android, and Amazon Kindle, to mention very few.

In addition to the value chains constantly morphing as the markets converge, it is interesting to note that different elements of the value chain change and evolve at different speeds. For example, the pace of innovation at the application areas is accelerating as barriers to entry for application developers are being lowered.[3] What about the rest of the value chain? In his book, "Clock Speed," [FIN], Charles Fine says "we are only in the dawn of understanding the second law of supply chain dynamics: *As you move closer to the customer in the supply chain, the clockspeed increases, sometimes dramatically.*"[4] By "clock speed," Fine refers to the concept he introduced to measure the rate of change of an industry or a product. For example, in the car industry, the median life of a car is 9.2 years and the car design cycle is 3.5 years. In contrast, consumer electronics have a median life of 2 years and a design cycle of 6 months. So according to Fine, and from what we can observe, the rate of change of elements of the value chain increases as you move up the chain and closer to the customer/user.

[2]iPhone owners have already (at the time of writing) downloaded 60 million applications from Apple's App Store.

[3]One VC firm, iFund which runs a $100M fund dedicated to start-ups that develop potential iPhone applications has received 2,500 business plans in the last several months. They have invested in four of these companies.

[4]p. 97.

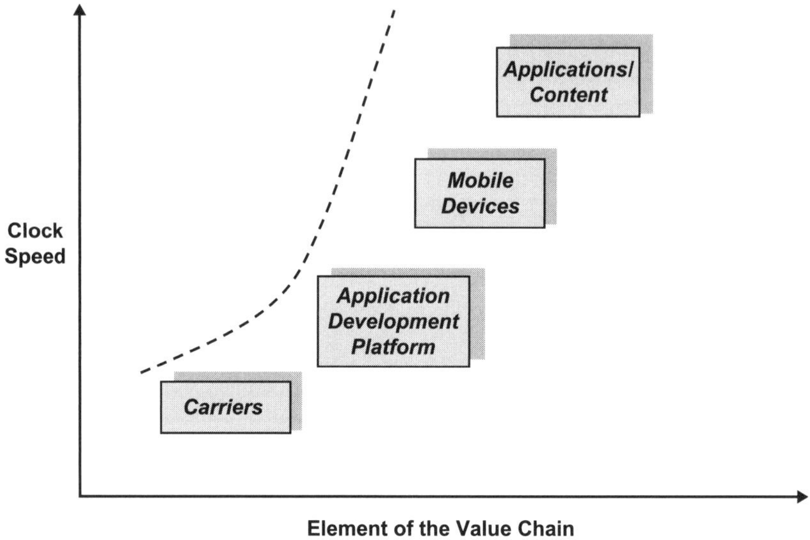

Figure 2-2: "Clock speed" of a value chain illustrating the "second law" of supply chain dynamics for generic mobile services.

Platforms change slower than the applications running on them – obviously and by definition. Core technologies that support the platform change slower than the platforms. This all ties back to the fundamental of modular design and the nature of economies of scale as scope [BAL]. Figure 2-2 illustrates the "second law" of supply chain dynamics for the generic mobile service.

An additional factor that complicates the market dynamics is the "Network Effect." At times, new and innovative mobile services enter the market offered by startups and fueled by venture capital money. These services are often offered for free and without a sustainable business model. The hope of many new application providers is that their service benefits from the network effect, where its value increases significantly with the number of people using it [BEN].[5] Furthermore, they also hope that the service is viral in terms of its adoption curve and that it will reach a tipping point quickly [GLA]. The endgame for such companies that do not have a sustainable way to derive revenue from their offerings is to be bought by a larger player for their customer base, or intellectual property, or both. Such merger and acquisition activities (particularly ones involving larger players) disrupt and then sometimes stagnate innovation in the value chain, as there are fewer players with fewer business objectives.

[5]In the narrow usage, a network effect is a characteristic that causes a good or service to have a value to a potential customer, which depends on the number of other customers who own the good or are users of the service. In other words, the number of prior adopters is a term in the value available to the next adopter. One consequence of a network effect is that the purchase of a good by one individual indirectly benefits others who own the good, for example, by purchasing a telephone a person makes other telephones more useful. The resulting bandwagon effect is an example of a positive feedback loop.

Any player in today's complex mobile services market asks the following fundamental questions:

- How can I make best use of my unique market position to drive and benefit from mobile services including social networking?
- Which revenue models (such as network traffic revenue, subscriptions, advertising, and transactions) will dominate the mobile services market, how will they evolve, and how will they differ by type of service?
- What form of relationship do I need to create with other players and the end user for a mutually beneficial environment in which the greatest market opportunities are realized?

One of the apparent trends in today's mobile services market is increased transparency of the services to the users. Users are much more in control, with information available about content, services, and devices to make them more discerning in their choices. The utility of a mobile device has expanded from traditional voice communication to incorporate text messages, transfer multimedia content, gaming, and Internet access and also to capture pictures and record video. Furthermore, the availability and familiarity of camera handsets and digital photo sharing has brought in the "user-generated content" market from Web 2.0 into the mobile realm. In Figure 2-3, we show an example of the functional enablers that a mobile operator can offer in a wireless content delivery value chain.

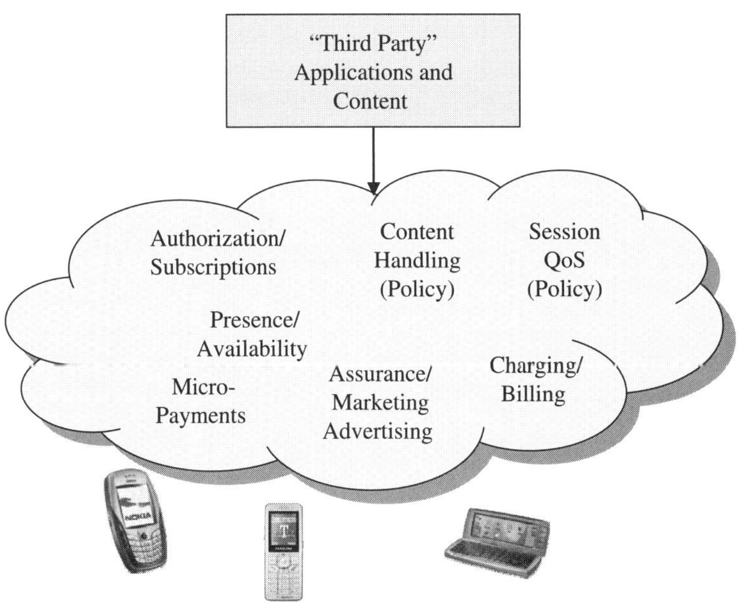

Figure 2-3: A content delivery value chain that involves mobile operators' services enablers.

Music Labels

EMI, Sony/BMG, Universal, Warner Music Group

Aggregators and Technology Enablers

Apple, Openwave, Real, Napster, Groove

Mobile Devices

LG, Nokia, Motorola, RIM, Apple, Sony/Ericsson, Samsung

Carriers

Verizon Wireless, AT&T, Sprint T-Mobile

Figure 2-4: Snapshot of the wireless music value-chain (adapted from [MCC]).

In Figure 2-4, we sketch out some of the players in the Music Delivery value chain. In this value chain, we find Apple both as an aggregator (via iTunes) and as a provider of a mobile device (iPhone).[6] The major and independent recording labels license the rights to their artists' recordings to music content aggregators such as eMusic, iTunes, Napster, and RealNetworks, which are typically classified as "music stores." Once a licensing agreement has been established, the aggregator converts music into a digital format that is accessible through either the mobile device's deck (i.e., "on-deck distribution") or through side-loading (downloading from a user's PC to the handset). To date, carriers appear to be split in their desire to own the music store (e.g., Verizon Wireless) or partner with existing online music stores to attract their subscribers to the carrier's music service (e.g., AT&T).

Due to the breadth of ways of obtaining music and the inherent convenience of using a phone as a music player, wireless music has experienced few difficulties in piquing consumer interest and stimulating demand. Content is still king as premium content sources hold key market leverage and great power over "user experience." Traditionally in the telecom world, end user experience was fully controlled by the phone company. Nowadays, you can activate your iPhone using iTunes, which is a result of an agreement between Apple and AT&T (provisioning interfaces and so on). Now imagine that Disney would like to deliver wireless content in a different way than Apple – now AT&T will have to agree on a different set of interfaces. To its credit, the telecom industry, through the TeleManagement Forum, has been working on common business process, information and application models, and plug-and-play standards. These types of rules are generally optional within a wireless mobile operator's infrastructure, but when they want to connect externally, they need standards to make things happen. The telecommunications industry has 100 years of experience with standards (which have led to modern standards such as Internet

[6]Given the ability to provide services as a Mobile Virtual Network Operator (MVNO), we may see, if the business conditions are right, further "bundling" of the layers (as well as possible unbundling).

Protocol), which were born out of a need to connect things together. For example, the TM Forum is involved in developing an extensible framework to deal with managing context, devices, and services. The content and Web worlds, however, may not share the same view.[7]

2.2 BETWEEN TWO WORLDS

Mobile services today exist in a duality. On the one hand, mobile devices can be viewed as an extension to the long-standing telecom service and the mobile applications as extensions to the phone company services. On the other hand, mobile devices can be viewed as extensions to the personal computer; devices that just happen to support voice communication. It is difficult to straddle both of these worlds simultaneously since they reflect very different mindsets. In the first mindset, we may feel, perhaps out of habit, that the telecommunications network operator is in the driver's seat, allowing or disallowing certain features on the device or certain applications. In the second mindset, we feel that we own the device and the information on it and are intolerant of any restrictions. In this section, as in the entire book, we try to speak about these two worlds in the same breath. After all, both are a manifestation of the same reality.

From our experience, "bell heads" and "Web heads" do not have much respect for each other. The first group believes that the second group does not understand issues of scalability, availability, and service level guarantees and gets too enamored with sexy applications and sizzling buzzwords. The second group believes that the first group "just does not get it." What they do not get is that it is a new world now and the intelligence embedded in the technology is so great that we do not need the stodgy methodologies of the old world of computing. This dichotomy, most likely, reflects some fundamentals of human nature and is not necessarily unique to technology, in general, and to mobile services in particular. In the rest of this chapter, and elsewhere in the book, we will look at aspects of the business landscape from both perspectives. By doing so, we are running the risk of alienating members of both groups, but hoping that perhaps by putting these apparently mutually exclusive worlds in such close proximity, we will benefit both.

2.2.1 Cooperation vs. Competition

In the previous section, we examined the complexities of the value chain in some specific areas. It was clear that the value chain is not linear and that cross-connections between actors exist almost in an arbitrary pattern. A company that offers mobile devices (e.g., Nokia) may buy a company that offers maps (e.g., Navteq) or a company

[7]The Web already has a set of standards (e.g., the IETF and the W3C) that may be sufficient. Also, the internet/Web protocols are easier to extend as they are text-based (ASCII) protocols. The traditional telecommunication protocols are binary (ASN.1) protocols that involve inflexible bit manipulations and are very hard to extend. Which approach will win?

that builds mobile operating systems (e.g., Symbian), and the value chain gets warped. It has been recognized for several decades now [BRA] that value chains are no longer "well behaved" and that relationships between the players can no longer be thought of as either "cooperation" or "competition" (i.e., only one or the other). The success of most businesses is dependent on the success of others, yet they must compete to capture value created in the market and protect their own interests. Therefore, business strategy and timing become key and impact alliances as well as merger and acquisition activities (see also [CON]).

A result of the constant change is a highly competitive environments where rivals can emerge overnight from unexpected places, such as traditionally noncompeting industries or from a lower segment of the value chain. In his seminal book, *Only the Paranoid Survive* [GRO], Andrew Grove said: "We live in an age in which the pace of technological change is pulsating ever faster, causing waves that spread outward toward all industries. This increased rate of change will have an impact on you, no matter what you do for a living. It will bring new competition from new ways of doing things, from corners that you don't expect." Changing business dynamics make collaborations more necessary than ever.

The model of cooperation between companies that are otherwise competitors to create maximum value is one of the most important business perspectives of recent years. Internet and mobile technologies have made it even more necessary for companies to both co-operate and compete by enabling relationships through information sharing as well as integrating and streamlining processes. Business is no longer viewed as a "winner takes all" or "zero-sum" game. The networked economy moves away from these purely competitive plays to recognize co-operative relationships that leverage value created by those in the network. In today's networked economy collaborations are a powerful means of identifying new market opportunities and developing business strategy. Competition then raises it head after the businesses have created value in the market and seek to allocate market share, price, cost, and other finite benefits. The management science of making sure the cooperative relationship is productive has become an art.

Competition usually leads to proprietary solutions that lock out alternative suppliers. Cooperation, to the contrary, can lead to open systems that may have many levels of published interfaces, allowing suppliers to apply their solution at any level, some times with loss of efficiency. In cooperative competition, the art is to find balance between a certain level of openness without sacrificing efficiency, and allowing competition in providing solutions that still adhere to the standards and interfaces that come with openness.

In his book *The Gridlock Economy: How Too Much Ownership Wrecks Markets, Stops Innovation, and Costs Lives*, Michael Heller [HEL] states: "In the telecom kingdom, today's great wireless knights – AT&T, Sprint, T-Mobile, and Verizon – often must license thousands of patents to provide any one consumer product. Even in existing markets, if they miss a single high-tech horseshoe nail, a whole network may be threatened. Worse, the fear of missing one patent defers investment. Whether a new technology ever infringes any other patent is never clear. At any moment, wireless providers and their suppliers are involved as plaintiffs, defendants, or interested third parties in a myriad of lawsuits." A case in point is the lawsuit against Research in Motion for patent infringement that threatened to shut down the BlackBerry service. To summarize, with the highly fragmented mobile services market, cooperation is more important than ever, and intellectual property battles threaten to deprive end users of value-added services. In the next subsection we will look at a line of battle between the world of the telcos and the world of the Web. It focuses on the role that telecommunication services providers play in the value chain and the ecosystem.

2.2.2 Dumb vs. Smart "Pipes"

The term "dumb pipe" denotes the telecom network that transfer bits without knowing anything about the application these bits comprise.[8] This term was coined by people in Silicon Valley who view traditional telecom networks as stodgy and slow to innovate, and who do not own networks! From the business side, being a dumb pipe may not be half-bad as the carriers make lots of money by providing access to their network. Furthermore, the carriers are increasingly asked to provide services such as security, location, and billing, which are essential to any realistic service offered on the mobile network. Traditional thinking pits the established network providers against the emerging Web industry of service providers such as Yahoo, Google, and Amazon, and views the struggle as an "either or" game in which network operators are likely to become dumb pipes. While this is one realistic outcome, there exist other business models for both parties, if they understand where they respectively add the most value in a value chain. This is the model of the "smart pipe," in which a network provides transactional and contextual value that enhances services – even to those services provided by third parties.

During the Bell System monopoly of the wireline telecommunications services, services were all bundled. One company was providing all network services – telephone services, telephone equipment, directory assistance and various "back office" operations including provisioning of new services, and billing. Then, in the late 1980s and through the 1990s, the telecom industry defined and adopted standards for the "Intelligent Network" (IN). The objective was to make it easy to create and deploy new applications and new services and overcome the closed, "switch centric" nature of deploying new services. Plain old telephone service (POTS) is a term that describes the voice-grade, basic telephone service that remains the basic form of residential and small business service connection to the telephone network in most parts of the world. The name is a reflection of the telephone service still available after the advent of more advanced forms of telephony such as mobile phones and VoIP. It has been available almost since the introduction of the public telephone system in the late 19th century, in a form mostly unchanged to the normal user despite the introduction of Touch-Tone dialing, electronic telephone exchanges, and fiber-optic communication into the public switched telephone network (PSTN).

[8]Another popular analogy is that of a "piano mover" that does not know how to play music.

The Intelligent Network (IN) is a network architecture that is intended both for fixed as well as mobile telecommunication networks. It allows operators to differentiate themselves by providing value-added services in addition to the standard telecom services. IN provided a more flexible way of adding sophisticated services to the existing network. Before IN was developed, all new features and services had to be implemented directly in the core switch systems by the switching equipment manufacturers. This made for very long release cycles as the bug hunting and testing had to be extensive and thorough to prevent the network from failing. With the advent of IN, most of the new services, such as toll-free numbers and geographical number portability, were moved out of the core switch systems and into self-serving nodes, thus creating a modular and more secure network that allowed the services providers themselves to develop variations and value-added services to their network without submitting a request to the core switch manufacturer and waiting for the long development process. The initial use of IN technology was for number translation services, for example, when translating toll-free numbers to regular PSTN numbers. Examples of such services are toll-free calls such as 800 number calls, call screening, call waiting, call forwarding, prepaid calling, and telephone number portability. In isolation these services may seem simple and of limited use, but when put into interaction complex telephony services could be built.

While the Intelligent Network did not foster an abundance of new applications, it did support massive scaling of few applications, for example, 800 numbers, roaming (the basis for mobile telephony), voicemail, and prepaid. The first problem that IN had was the business model. Intelligent Network functionality is service-affecting and therefore carefully controlled. Outside of a few large equipment suppliers, developers do not get access to the IN. As a result, most new services were invented at the edge. For example, voicemail originated as an enterprise application and was only integrated into the network many years later. Even prepaid phone service was implemented from the edge and only integrated into the IN over a period of years.

Wireless carriers now have the opportunity to engage in profitable value chain partnerships (involving revenue sharing and transaction fees) with Web companies, and content and e-commerce providers, based in the exchange of real-time "user information" (privileged information such as mobile user identity, security, credential, micro billing, handset presence, service and feature entitlements, location, demographics, etc.) using the IMS Service Delivery Platforms (SDP) as discussed in Chapter 1. Besides convergence economies (all services on IP), the objective of IMS is to facilitate the rapid development and deployment of new applications and new services. IMS may be following exactly the same path as IN [TUR]. The network-based SIP application servers that define IMS call flows are critical to basic telephony. As a result, no operator lets third-party developers run applications on their IMS core. Even worse, most operators maintain a walled garden for data access.

Another key issue is service creation software. Unfortunately for both the IN and IMS, the standards focus on basic call control and media flows; that is, on the plumbing. There are no IMS standards for software APIs or software development environments. Typically, each major equipment provider has a set of service creation tools, but they differ from vendor to vendor. The wireless value chain is progressively disintermediating at the "application-device" and "service control-network" boundaries, with ongoing deployment of XML/Web Services and SIP based infrastructure and service convergence technologies.

2.2.3 Opposing Forces at the User Level

This subsection looks at some of the seemingly contradictory opposing forces that face the entities in the mobile services chain.

Complexity vs. Simplicity As the networks and application grow in sophistication, they also threaten to overwhelm the end user. Achieving simplicity, however, is not easy; (see, for example [NOR] [NOR2]) poor design and usability choices at prototype and development phases may mar the whole lifecycle. We discuss this issue at length in Chapter 4.

Expecting Predictability vs. Embracing Irrationality When introducing a new mobile service (or any consumer product), how can we know in advance if it is going to be adopted? The literature is full of books that discuss methodologies for testing user preferences and predicting user behavior [AND1]. However, no matter how sophisticated our consumer research models are, people always catch us by surprise. The reason may be lying in our relying on standard economics when perhaps a more seemingly irrational economics, called behavior economics, is at play [ARI]. Standard economics assumes that all human decisions are rational and informed, motivated by an accurate knowledge of the value of all goods and services and the amount of happiness (utility) all decisions are likely to produce. Under this assumption, everyone in the marketplace is trying to maximize profit and optimize his or hers experiences. Behavioral economics, on the other hand, people are susceptible to irrelevant influences from their immediate environment, which is called "the context effect," irrelevant emotions, shortsightedness, and other forms of irrationality. These mistakes also provide opportunities for improvement. If we all make systematic mistakes, we can devise tools and strategies to counter them.

2.3 INNOVATIVE BUSINESS MODELS

In the following section, we will examine a few innovative business models that deviate from the traditional telecommunication/content business arrangements. The key common element here is the unbundling of the traditional offerings where the provider of the telecom service owned a network, or where the content downloaded over the network was tied to a telecom service provider or to a device and location, or where the recipient of the service or the content is the one paying for it. These models represent a trend where business boundaries shift as the value chain is rearranged.

2.3.1 The MVNOs

The Mobile Virtual Network Operators (MVNOs) do not own wireless spectrum or network infrastructure. This allows them to assume less risk, lower upfront costs, and reduce the time it takes them to start operations. Usually it takes an MVNO about three months to be up and running. MVNOs resell services from network-based mobile operators under their own brand – in this way they "own" their customers! They differentiate themselves through specialized content, services, or handsets, target under-penetrated markets and niche segments such as youth, credit-challenged individuals, families, ethnic groups, and sports fans. MVNOs currently "own" a few percent of the worldwide mobile users market. The value chain consists of three major players (Fig. 2-5):

1. MVNOs offer strong brand identities and segmented market focus; examples include Virgin Mobile and Movida.
2. Mobile Virtual Network Enablers (MVNEs) offer the back office, logistics, and services to support the MNO/MVNO relationship; examples include Telcordia Technologies and Verizon.
3. Mobile Network Operators (MNOs) utilize excess capacity and eliminate high customer care/support costs by selling wholesale services to the MVNE/MVNO; examples include Sprint, Verizon, and AT&T.

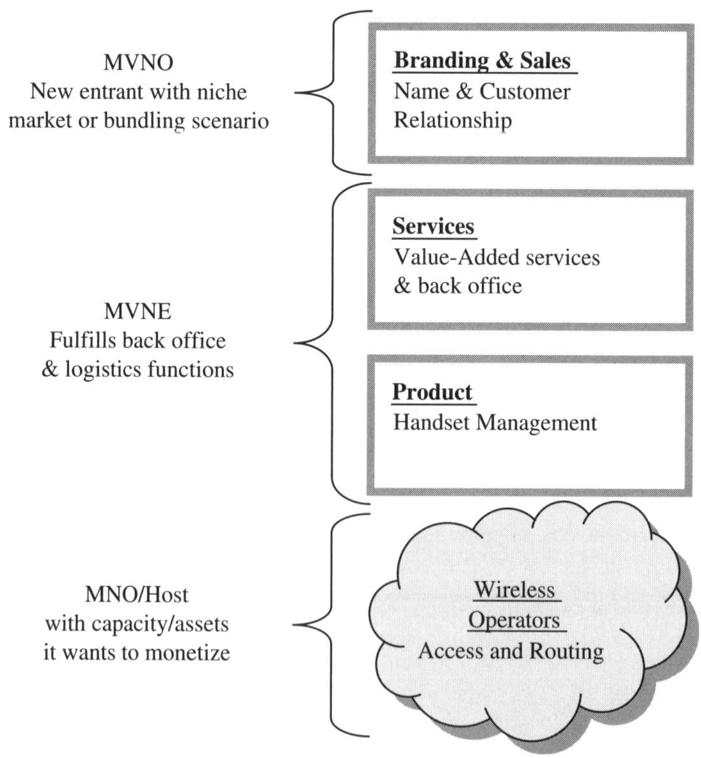

Figure 2-5: One example of the MVNO Value Chain.

The MVNO business model/value chain illustrates how business boundaries can move freely along the value chain to fit the need of the service provider. For example, in Figure 2-5, the boundaries separating the businesses of the MNO, MVNE, and MVNO can move up or down. In some cases, the MNO may be offering and managing the handset, while in others it may be the MVNE. Also, the role and responsibilities of the MVNO may sometime include back office functions. The beauty of this emerging business model is its flexibility and the fact that functions and roles can be allocated between the various providers based on business context and customer needs.

2.3.2 Communications as a Feature

In late 2007, Amazon introduced the Kindle, which is a device that allows you to read books that you download from the Amazon store for a one-time fee per book. Some of the most important features of Kindle are listed below (Source: product and service descriptions on Amazon.com website in September 2008):

- Revolutionary *electronic-paper* display provides a sharp, high-resolution screen that looks and reads like real paper.
- Simple to use: no computer, no cables, no syncing.
- Wireless connectivity enables you to shop the Kindle Store directly from your Kindle – whether you're in the back of a taxi, at the airport, or in bed.
- Buy a book and it is auto-delivered wirelessly in less than one minute.
- Lighter and thinner than a typical paperback; weighs only 10.3 ounces.
- Long battery life. Leave wireless on and recharge every other day. Turn wireless off and read for a week or more before recharging. Fully recharges in 2 hours.
- Unlike WiFi, Kindle utilizes the same high-speed data network (EVDO) as advanced cell phones – so you never have to locate a hotspot.
- No monthly wireless bills, service plans, or commitments – we take care of the wireless delivery so you can simply click, buy, and read.
- Email your Word documents and pictures (.JPG, .GIF, .BMP, .PNG) to Kindle for easy on-the-go viewing.

One of the unique aspects of the Kindle as a device supporting a mobile service is the fact that "communication" is not the main aspect of the device – it is only a feature. This signifies a trend in feature rich smart devices. In these devices, communication may be a feature that is not always visible to the user both at the technical level and at the service level. For example, in the case of the Kindle, connectivity to the Amazon.com portal is provided by Sprint but the user is not aware of the telecommunications provider and does not have to pay a separate phone bill (part of the price of content subsidizes the connectivity costs). A natural question that arises is why anyone should buy a Kindle device when one can use the iPhone or other smart phone, equipped with the appropriate software application, to read books? Why buy a specialized device when a general-purpose mobile device can do the same job?

Table 2-1: Comparison between the Kindle and the iPhone

	Kindle	iPhone
Wireless Service	Always-on, no subscription fee, Sprint EVDO (value $50/month). Connection is part of the sales price of the content. No need for Wi-Fi hotspots.	Monthly service fees for the iPhone data service. If a book reader application is used (e.g., Stanza) fees for the book are additional.
Mobile Device	Special purpose point-of-sale device that is connected only to the Amazon.com store, which sells a large variety of merchandise in addition to books.	General-purpose mobile device, but connected to the Apple iTunes store.
Some Features	The Kindle has a hardware keyboard, does not have a camera, is not designed for human-to-human communication. Supports HTML, PDF, and Word through document email system. Currently, no management for collections of documents.	No hardware keyboard, has camera, fits in your pocket. Designed for communication. The user can watch TV and even read books (say using the Stanza application). Supports HTML, PDF, Word, and PowerPoint formats. Currently, no management for collections of documents.
App. Developers Ecosystem	No such ecosystem for applications for the Kindle. The Kindle promotes a content ecosystem but Amazon does not benefit financially from it.	App Store creates an incentive for software developer to build applications for the iPhone. Apple keeps a percentage of the sales of the applications.

Source: Adapted from [JAN]

In Table 2-1, we examine some of the pros and cons of the Kindle vs. the iPhone running eBook software. This discussion is based on information provided in [JAN].

Another example of an interesting unbundling of traditional business constraints is the SlingBox, which is offered by Sling Media.[9] The Slingbox is a TV streaming device that enables users to view their cable, satellite, or personal video recorder programming from an internet-enabled computer with a broadband internet connection (known as "place-shifting" and "time-shifting"). Slingbox can redirect up to four live audio/video signals to anywhere in the world, but only one user can access the stream at a time (due to copyright issues).

2.3.3 The Economy of Free

The business and technology ecosystem of mobile services is complex and highly interconnected, with constantly shifting business boundaries and technology

[9] www.slingmedia.com.

capabilities. One of the trends we have not mentioned much above is that of providing services free to the end user. How can a business sustain itself while offering at least some of its services free? The quote at the beginning of the chapter points out the fundamental difference between "free" and "1 cent" [AND3]. If a business manages to remove the complexities, as illustrated by the use case above, and manages to offer a service for free, life is good (e.g., Google). Table 2-2 highlights some aspects of the "free" business model.

The ideas described in this section are thought provoking and, in order to get a sense for how they could play out, we will provide in the following a somewhat detailed example of offering a personalized mobile service and the business decisions confronting the service designer and provider in today's market realities. The example illustrates that it is not easy to put the general observation we made in this chapter into practice.

Table 2-2: The economy of free

	What is free?	Free to whom?	Who makes money?	Who pays?
Advertising	Content, services, software, and more	Everyone	The advertising platform provider, the content, service, software provider	The company that advertises, the content, service, software provider
Cross Subsidies	Any product that entices you to pay for something else	Everyone who is willing to pay eventually one way or another	The product provider	The product provider and the end user (everyone)
Zero Marginal Cost	Things that can be distributed without an appreciable cost to anyone (e.g., music)	Everyone	No one	No one
Labor Exchange	Websites and services	All users, since the act of using these sites and services actually creates something of value	No one	No one
Gift Economy	Open source software or user-generated content	Everyone	No one	No one

Source: Adapted from [AND3]

2.4 CASE STUDY: PERSONALIZED NAVIGATION DEVICES

In order to illustrate the dynamic and even volatile nature of today's market place and technology and network road map, we will analyze in the following the case of offering a personalized mobile service. Despite the enormous potential of personalized mobile services, their success depends on overcoming several business challenges. We will discuss some of these challenges in the context of connected personal navigation devices (connected PNDs). The use of connected PNDs will increase in the future and these devices will capture a significant portion of the market.[10]

There are two main categories of connected PNDs. The first includes PNDs that receive a predefined list of content via a broadcast medium such as FM radio (e.g., RDS-TMC, which stands for Radio Data System Traffic Message Channel), satellite (e.g., SIRIUS Travel Link), and Digital Video Broadcasting (DVB). The second includes PNDs that are able to connect via one or more communication technologies to the Internet or a data center owned by the provider of the device and retrieve relevant information, such as traffic alerts, movie listings, and local business listings, in real-time. The following list enumerates some of the business problems associated with connected PNDs that receive information via a broadcast medium.

- *Broadcast Technology* – RDS-TMC technology broadcasts near real-time traffic information that can be used for both informational purposes and alternate route planning. Satellite broadcasting typically includes a broader spectrum of information, such as traffic and weather information, movie listings and play times, gas prices, and sports scores. The location of the PND (GPS coordinates) is used as the filtering mechanism for selecting only information relevant to the current location. While satellite broadcasting seems to be a natural choice due to its wider range of available information, the limited number of available providers and chipsets requires appropriate business arrangements and partnerships to be put in place (e.g., Ford SYNC and SIRIUS Travel Link). In some cases, exclusivity agreements may need to be established in order to offer an advantage to one or more of the involved parties.
- *Built-in or Accessory* – Broadcast data services require dedicated receivers. These receivers could be integrated with the PND and included in the same enclosure or be accessory components that connect with the PDN via a wired or wireless connection. For instance, TomTom PNDs support an accessory RDS-TMC receiver,[11] while the Ford SYNC approach follows the integrated approach and bundles the SIRIUS Travel Link with the on-board navigation system. Deciding which approach to follow depends on several factors, including partnership arrangement with data providers and pricing models.
- *Broadcast Costs* – Depending on the selected broadcast technology and the country in which the technology is used, different pricing options may be

[10]See, for example, the Telematics Research Groups reports in www.trg.com.

[11]http://www.tomtom.com/page/rds-tmc

available. For example, RDS-TMC services in Europe offer one-off cost for lifetime subscriptions while similar services in United States require yearly subscriptions. Satellite-based information services may offer monthly subscriptions and require additional satellite services. For example, SIRIUS Travel Link requires SIRIUS audio subscription in Ford SYNC. The primary business decision that has to be made here is whether to offer a one-off price for the device and the services bundled with the device or charge for the device and then offer a monthly or yearly subscription fee that depends on the services selected by consumers.

Connecting the PND to the Internet or the data center could be an expensive proposition. Table 2-3 enumerates some of the technical decisions that will considerably affect the pricing of the service.

Here are additional details:

- *Connectivity Method*: A PND can connect to the Internet utilizing cellular, Wi-Fi, or both. When connectivity is achieved using cellular technologies, such as those offered by the CDMA and GSM family of protocols, access to Internet content can be accomplished using one of the following communication channels: SMS, voice, or data. When the SMS channel is used, SMS messages are used for sending and receiving relevant information. When the voice channel is used, modem or data-over-voice technology could be used for content access. When the data channel is used, HTTP, TCP/IP, or UDP/IP can be used for Internet connectivity. Depending on the functionality of the offered services, selecting which cellular channel to use is an important business decision that affects both pricing and service functionality.
- *Built-in versus External*: Selecting whether to offer built-in connectivity or rely on another device for doing so is another important business decision. Several existing PNDs support Bluetooth connectivity with cellular phones and can use the cellular phone for sending and receiving information. For example, Dash Express, from Dash Navigation Inc. offers built-in cellular and Wi-Fi connectivity for the purpose of sending to and receiving from the Dash Network Operations Center real-time traffic information for the purpose of selecting the least congested route to a destination. A built-in solution requires carrier

Table 2-3: Issues in providing the PND with connectivity

Issue	Options
Connectivity method	Cellular, Wi-Fi, or both (SMS, voice or data).
Connectivity device	Built into the PND device or provided externally (e.g., a cell phone connected via Bluetooth).
Pricing model	Price per use or flat monthly or yearly rate.
Location data sharing	Negotiate agreements with partners when user opts in.
Partner selection	Select partners to offer differentiated content through portal.

selection and plan negotiations, a tedious and lengthy process. An external solution may impose requirements on the service plan available on the device used for communications (e.g., unlimited data plan). A major advantage of the built-in solution is that it can function in a stand-alone mode. An external solution, on the other hand, requires a cellular phone or another communications device and network services are available only when a communications device is available and paired with the PND.

- *Pricing Model*: There are two main alternatives to be considered. First, a pay-per-use model can be used that charges consumers when they connect to the Internet and access information. This is similar to the model used by Amazon's Kindle electronic book reader device as mentioned above, or the iTune model, where the consumer is charged a fixed price for the download of a book or some other electronic content. Second, a fixed service price per month or year model can be used, similar to existing cellular service plans. While offering great flexibility in terms of pricing plans is desirable, offering too many options may be confusing and result in dissatisfied consumers.
- *Location Data Sharing*: Connected PNDs allow providers to collect valuable location data about their subscribers. The collected data present a wealth of information that could be shared with partners. However, reaching agreements on how to share these data and which parts of the data to share can be quite complex and time consuming due the various regulatory requirements around privacy.
- *Partner Selection*: To offer better services, it makes sense to provide a portal that enables consumers to access content and information relevant to them from a central location. This makes it easier compared to having consumers search the Web for such information. Obviously, the content and information available at the portal plays a critical role in service adoption. Selecting the right partners to offer such information is therefore an important business decision.

This use case provided some concrete illustrations of how the various business decisions play out when one confronts the need to commit to a specific service with a specific architecture in the face of extreme market uncertainties.

2.5 SUMMARY

The ability to enable communication over great distances between mobile users introduces a profound shift in the nature of not only social and political relationships, but of commercial ones as well. New business models have evolved around the use of communications over distances to sell and trade and to enhance customer intimacy and care. The possibility for offering new services targeted to mobile users has captured the imagination of marketers and technologists alike. Location-based services (LBSs) have since become a catch-all term that includes any service offered through a wireless handset that utilizes positioning technology (such as GPS) to pinpoint a user's location and offer location-centric value to the user. In the next chapter, we examine many aspects of location as an enabler for mobile services and for the mobile experience.

QUESTIONS

1. If mobile handset manufactures were to merge with telecom providers, how would the ecosystem be affected?
2. If Wi-Fi or some other wireless technology were to become ubiquitous, what would the role of a telecom provider be?
3. What happens if you have a mobile reading device that includes wireless connectivity to a portal through a third-party network operator and that operator goes out of business or changes direction? What are the legal and user ramifications, for that matter, if any link of the value chain "breaks"?

3

MOBILE USER LOCATION AS A SERVICES ENABLER

"The best time is always right now; the best place is always right here."
—Greg Hickman

Technology enablers are software or hardware components that offer functionality that allows other entities to achieve something. For example, a Wiki is a technology enabler that allows information dissemination within an organization. Enablers can vary in size, scope, and scale, both in terms of the technology they encapsulate and in terms of their business role in the value chain leading to the user. As discussed in the previous chapter, enablers are an integral part of a "value chain" and act as building blocks for higher levels of the chain. Any specific enabler may be composed of other enablers that reside lower in the chain. Typically, service developers use "service creation environments" for "gluing together" service enablers. In the context of mobile services, there are several important sources of enablers. Open APIs to community services (such as social networks and photo sharing Web sites) can be thought of as enablers at the information level. In the mobile network infrastructure, authentication, authorization, and accounting (AAA) servers, billing servers, and registrars enable network services.

As a general point, enablers could be used by various other entities in many ways. For instance, the uLocate WHERE platform is a widget development platform that allows location-based application developers a quick path to market and, thus, it can be considered an enabler. The platform's API provides hooks into local search, GPS coordinates of the local device, and other geospatial functions such as

The Fabric of Mobile Services Software Paradigms and Business Demands, by Shoshana Loeb, Benjamin Falchuk, and Euthimios Panagos

computing distances between points. In the mobile media and entertainment market value chains, telecom service enablers include connectivity, billing (including real-time billing), security and authentication, location information, presence information, session/call control, policy functions, and, in some cases, user profile information. In the next generation (all IP) services value chain, IMS implements a set of shared "micro services" such as handset presence, location, single sign-on, user preferences, allowable services features, and more. These IMS micro services are services enablers designed for use by Service Oriented Architectures (SOA) and Web Services based Services Delivery Platforms (SDPs) for the creation and delivery of rich IP-based voice and multimedia services. To the Open Mobile Alliance (OMA), an enabler is an interoperable entity that enables interactions between different components and applications developed by different providers, such as device and network suppliers, information technology companies, and content and service providers. Here, enablers are collections of specifications that literally enable the realization of interoperable end-to-end functions in the mobile environment. In the Mobile Virtual Network Operator (MVNO) market value chain, services are created from enablers provided by the Mobile Virtual Network Enabler (MVNE). Examples of such enablers include billing, network element provisioning, handset activation, and operations support systems. Enablers can span all aspects of mobile services and, hence, it is not possible to cover them all in detail. The rest of this chapter is an in-depth examination of location enablers.

3.1 INTRODUCTION

"Where are you?" This simple question is asked over and over again when talking to friends, relatives, family members, or acquaintances using mobile devices. Humans need to know where others are and what they are doing right now. The same need exists for accessing contextual information related to a specific location and interacting with such information. It is clear that location plays an important role in our daily lives due to its unique ability to connect our surrounding physical world with digital information and content available for this physical world. Mobile devices are poised to become the primary delivery medium for a new generation of location-aware services.

The following excerpt from the May 2005 article, "The Geospatial Web: A Call to Action," by Mike Liebhold[1] paints a future where everything, from people, places, digital media, to emergency events, vehicle condition, phone calls, and other interactions, could have a location associated with it.

> Out of this teeming ecosystem, we can see the beginning shapes of a true geospatial web, inhabited by spatially tagged hypermedia as well as digital map geodata. Google Maps is just one more layer among all the invisible cartographic attributes and user annotations on every centimeter of a place and attached to every physical thing, visible and useful, in context, on low-cost, easy-to-use mobile devices.

[1] Available online at: http://www.oreillynet.com/pub/a/network/2005/05/10/geospatialweb.html

Being able to associate location information with a particular entity (referred to as geo-tagging), be it mobile, stationary, transient, or persistent, is just the very first problem that has to be addressed by services whose utility depends on location data, referred to as location-based services (LBSs). Once the location of an entity can be determined, many additional technical and business issues would have to be addressed as well in order for LBSs to be embraced by users. The following list includes just a subset of these problems:

- *Access to Location-Specific Content and Information.* Today, only a small fraction of the available digital content and information is associated with geographic information (e.g., GPS coordinates, ZIP code, etc.), such as cartographic data, road maps, and weather maps. Furthermore, existing repositories that either include geo-tagged content (such as those storing images taken by digital cameras with built-in GPS receivers, or business directories) or provide user location warehousing functionality (e.g., Yahoo's Fire Eagle service[2]) address only a very small fraction of the entire universe of entities that are or could be geo-tagged. The lack of a "centralized" access point for geo-tagged content and information, such as a search engine, limits the amount of contextual information available to LBSs to only the information present in the geo-tagged repositories these services have access to, or are aware of.
- *Association of Relevant Contextual Information with a Location.* A geographic location may have numerous geo-tagged entities associated with it, such as nearby coffee shops, restaurants, movie theaters, and museums. Filtering the potentially large number of such entities in a way that results in a relatively small set of entities that are of interest to users of LBSs is not a trivial problem. Even when the user specifies filtering criteria in the form of search queries of category selections, the resulting list may still be quite large and result in the so-called "information overload." Therefore, the service should incorporate pertinent user context information, such as current activity or next calendar appointment, in order to filter these entries and present only those that are deemed to match the user's intent.
- *Privacy and Security.* Many of us are quite concerned with network services that have access to our current location or collect location related information about us (e.g., home address, work address, and typical route between home and work). How can we be assured that our location information will not be disclosed to third parties and applications for marketing or other purposes without our explicit approval? Worse yet, how can we be assured that our location information will not be exploited for malicious or criminal purposes? Furthermore, how can the service provide reasonable guarantees that collected location information will be secured against unauthorized accesses? The above questions identify just a subset of the privacy and security issues that need to be addressed by LBSs.

[2]Accessible at: http://fireeagle.yahoo.net/

- *Service Charges.* Charging for services is always a very challenging business problem. Which service charging model should be used for LBSs? Would a monthly or yearly subscription approach work the best, or should a pay-per-use approach be offered? What about a service sponsorship approach where advertisers may either partially or fully pay for the service in exchange for including specific content in response to location-specific user queries? Or what about an approach where businesses pay for such services when users perform transactions with them in response to contextual information presented to users about these businesses and their offerings? The above questions demonstrate how complex the billing component of LBSs may be.
- *Integration with External Content and Services.* Stand-alone LBSs, such as those provided by the majority of existing personal navigation services, are quite limited in the functionality they provide. For example, a personal navigation service that does not have access to real-time traffic information will most likely "tell" us to take the same route from point A to point B independent of whether there is a road closure due to an accident along the route or not. Being able to integrate external content and services that are relevant to a specific location or geographic region is, therefore, an important property of user-centric LBSs. However, such integration comes with its own technical and business issues. On the technical side, integration with proprietary protocols may be required. In addition, external information may not follow a consistent geo-tagging approach or the geo-tagged information may not be at the same granularity as that required by the service. On the business side, selecting the "right" third party content and service providers is extremely important in incorporating relevant and timely content and information in LBSs.

While it is tempting to devote this chapter to discussing all of the above mentioned LBSs challenges, and even more, we will restrain ourselves and focus on a subset of these problems instead. In particular, we will discuss available mechanisms for calculating the location of an entity (the building block for being able to geo-tag entities); describe overall location-based system architectures, together with some important middleware components; touch upon privacy and charging alternatives; and explain the important role of search and information filtering in LBSs. Our hope is that these topics provide enough information to better understand the various components of LBSs and start thinking about new and exciting services that could be enabled by location information.

3.2 LBSs – THE KILLER ENABLER

Today, mobile handset processing power and functionality parallels that of laptop computers. In addition, handset manufacturers, platform/content providers, and wireless operators are making available Software Development Kits (SDKs) and libraries to enable third parties to offer applications for their handset or mobile platforms (e.g., Apple's iPhone, Open Handset Alliance's Android, Yahoo's OneConnect, Sprint's Titan).

In this new and very powerful mobile world, location-based services are gaining a prominent role due to their game-changing potential. According to ABI Research: "By 2013, we can expect to see more than 900 million GPS-enabled devices in the market; each offering an array of new and innovative location-based services beyond traditional navigation, such as automatic geo-tagging of pictures taken with digital cameras, road toll systems, and social networking applications."

However, widespread acceptance of LBS faces many technical, business, and psychological barriers. On the technical side, understanding the context in which mobile handsets are being used, the strengths and unique challenges of mobile handsets, and providing a simple and intuitive user interface are key factors in successful user adoption. On the business side, developing strategies for enabling advertisers, information, and content providers and making the service available to a wide audience is required for profitable services. On the psychological side, enforcing strict security and data lifecycle management will ease user worries associated with the privacy of the information collected and shared by location-aware services.

3.2.1 LBSs and Wireless Operators

Given the declining profitability in core voice business, wireless network operators have been searching for the "killer application" to drive enhanced services revenues for quite some time now. In the past, many vendors and industry analysts had hyped LBSs as the contender, but this positioning has distracted us from the true power of location. Today, a consensus is forming around a different view: LBSs are not a killer application but location itself is the "*killer enabler*" of existing and new wireless services. This consensus represents a sea change in the industry that will plot LBSs on a more successful course.

With advances in network infrastructure and next-generation handsets featuring large color displays, built-in cameras, increased memory, removable storage, and robust operating systems, together with business and consumer demand for enhanced services on the rise, LBSs are finally maturing from hype to something real, with compelling reasons for adoption. In this light, a strong value proposition can be constructed around LBSs that addresses two business challengers of the wireless network operator: 1) increasing average revenue per user (ARPU) and 2) reducing subscriber churn. Wireless network operators can achieve these objectives by both enhancing existing services with location and introducing new location-based services.

The easiest opportunity (i.e., "low-hanging fruit") for the wireless operators to monetize LBSs is to enhance existing voice and data services. Below, we will look at three existing services – directory assistance, roadside assistance, and messaging services – as examples of this trend. We have already discussed these examples somewhat in Chapter 1.

- *Enhanced Directory Assistance.* Also known as 411, directory assistance is one of the highest-margin services provided by wireless operators. Just as we discussed in Chapter 1 when we introduced the "personal assistant" service in its many levels of sophistication, the goal of location enhancement is to provide a more convenient service. Knowing the location of the user expedites service delivery while, at the same time, providing the potential to deliver value-added information services such as driving directions, points of interest searches, or current traffic conditions. Directory assistance is basically just 411 from the user's perspective, but LBS serves as its killer enabler, making it a more compelling and efficient service.

- *Enhanced Roadside Assistance.* Drivers expect roadside assistance to be location enabled. While this may indeed be the case for those that access roadside assistance as part of a telematics service, those with no means for retrieving their current location are less fortunate. Mobile operators are capable of retrieving the approximate location of a mobile phone (we will present details about how this is done shortly) and, hence, making this information available to roadside assistance entities. As we will discuss later in this book, sophisticated location-enhanced roadside assistance offers useful on-the-road services, such as providing distances and directions to gas stations, automobile repair services, or directions to nearby restaurants and hotels.
- *Enhanced Messaging Services.* Location is mostly irrelevant when participating in a chat room on the Internet or exchanging e-mails and instant messaging from a desktop computer. However, when people are chatting or text messaging with friends from their mobile phones, they want to interact with others who are nearby. To this end, "friend finders" and "mobile dating" are becoming very popular in Europe, as demonstrated by the success of location-enabled chat applications. With the addition of location to existing messaging services, daily short message service (SMS) traffic has soared, resulting in substantial increases in ARPU.

3.2.2 LBSs and Mass Market

Most LBS market predictions concur that personal navigation type services will continue to comprise a large part of global LBSs revenues. However, other location-based services are currently being offered for both mass consumer and enterprise markets. Here we outline some of the more interesting and relevant ones to our subsequent discussions. Interested readers should follow up at one of the many resources available on this topic such as LBSZone.com, ESRI's LBS portal (at ESRI.com), or NavTeq.com, which stages the related Annual Global LBS Challenge.

Personal Locator Services Personal locator services based on GPS-enabled mobile phones launched as "family-finder" options by Sprint Nextel, and Verizon Wireless as well as by the youth-oriented mobile virtual network operator (MVNO) Helio (acquired by Virgin Mobile, one of the very first MVNOs). The service allows users to build a "buddy list" and broadcast their locations to each other for display on GPS-linked maps. Despite their utility, however, personal locator services have some possible downsides as users must remember that they can switch the function on and off, and should manage their buddy lists prudently to avoid "stalkerware" implications.

A twist on friend finders is the "family minder" service that assists caregivers in keeping track of family members. Nothing, of course, can guarantee the safety of a child or an elderly relative or the ability to find a lost person. We will offer more information on this class of services in our discussion of geo-fencing below.

Geo-Fencing The term geo-fencing is used to indicate a violation of specific geographic boundaries when one either enters or exits these boundaries. For example, in the case of telematics services geo-fetching is used for notifying the owner of a vehicle when the vehicle travels beyond a specific geographic area. Today, many services offer geo-fencing functionality including telematics services and child-tracking services, such as the kids-focused MVNO Kajeet.[3] Child-tracking services allow certain friends and family members to access a Web interface to see the child's location. Parents can also set up "scheduled checks" to make sure their child is where they are supposed to be, when they are supposed to be there. Two examples of this service are:

1. AmberWatch Mobile offered via the AmberWatch Foundation,[4] a non-profit organization dedicated to the prevention of child abductions. The AmberWatch Foundation website reports that a child is missing every 40 seconds, and provides a ticker which shows approximately how many children have gone missing since you arrived at the site – a sobering statistic.
2. The Sprint Family Locator service allows parents to locate their child on a map through either a Website or through their phone. It requires parents to list up to four phones they want to be able to locate. Registration requires activation on both the "parent" phones as well as the child phones. When a parent wants to find a child, the child can be located on a map along with a street address and surrounding landmarks. One "cool" feature of Sprint Family Locator is that the child is notified by text message whenever a parent is snooping on them.

Similar services are offered to locate the elderly as well as to inform car insurance companies of teen driving behavior (with incentive of reduced premium for best behavior) and tracking services for pet through a mobile device in their collar.

Personal Navigation Services Throughout the book we will be providing plenty of examples that focus on personal navigation services that, as their name implies, provide turn-by-turn directions to get from point A to point B. Until recently, these services were not offered by a wireless network operator but were rather GPS enabled and running locally on a specialized device, known as Personal Navigation Device (PND).

[3]See http://www.kajeet.com for details.
[4]See www.amberwatchfoundation.org for details.

Dynamic Traffic Alerts In addition to familiar alerts such as news updates, sports scores, stock prices, and weather reports received on text-messaging-enabled mobile phones, many wireless network operators, telematics service providers, and information providers are beginning to offer location-enabled dynamic traffic alerts to their customers. In survey after survey, traffic information has always ranked highest among the data wireless subscribers want. With a typical service, drivers can specify their routes and when they would like to receive traffic updates. Travelers can then alter their routes and plans in response to real-time traffic reports for their particular route and schedule, saving themselves time, fuel and frustration.

Mobile Resources Management (MRM) With location-enabled mobile resource management (MRM) or field force automation, businesses can locate, communicate, and manage their mobile workforces in real time, reducing service response times, and improving customer satisfaction. Many wireless network operators, especially in Europe, have begun to recognize MRM as a key differentiating service they can offer to small- and medium-sized companies in a wide range of industries. MRM gives these business customers the same advantages as larger enterprises, but without requiring a capital investment in new IT infrastructure. MRM integrates seamlessly with existing networks and equipment, enabling these businesses to communicate with mobile workers through SMS on their handsets, and to produce reports on worker status for distribution to other business units.

3.3 POSITIONING TECHNOLOGY

Positioning technology refers to the specific mechanism used for determining the approximate position of an object, be it fixed or mobile. In the case of mobile devices, there are two main approaches to implementing positioning services: device-based and network-based. Under the device-based approach, software and hardware components associated with a mobile device are used for computing the approximate location of the device by either performing calculations at the device itself or utilizing network-based services for these calculations. Under the network-based approach, the mobile network and network-based position determination equipment are used for calculating the position of a mobile device. In the remainder of this section, we will provide an overview of the various techniques currently available in a variety of networks.

3.3.1 Cellular

The main components of a typical cellular network are: Mobile Stations (MSs), Base Stations (BSs), and Mobile Switching Centers (MSCs). MSs (e.g., cellular phones) communicate with BSs over radio. BSs are fixed structures (cellular antennas) that connect to MSCs. Each BS is responsible for serving a subsection of the cellular network coverage area. This subsection is referred to as cell. Each MSC communicates with several BSs. Figure 3-1 shows a typical cellular network with each cell being represented by a hexagon.

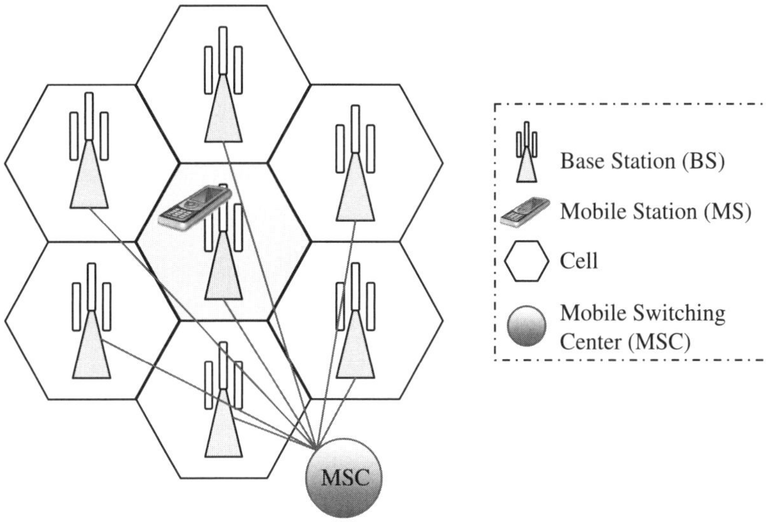

Figure 3-1: High-level Cellular Network Architecture.

Each cell served by a BS is assigned a unique ID. By knowing the exact locations (e.g., GPS coordinates) of BSs and the approximate sizes of the cells being served by them, one could estimate the position of a MS using the ID of the cell associated with the BS serving the MS. This method of positioning is referred to as *Cell ID* or *Call of Origin* (COO). While wireless providers maintain their own inventory of BSs and their exact locations, this information has not been readily available to third parties. Fortunately, however, a number of companies have started building such databases and making them available in recent years. These include CellDB, CellSpotting, and CellTrack, all of which make cell ID information available for subsequent exploitation in services.

Since MS positioning using Cell ID depends on the size of the cell, the accuracy of this method can be quite low (a single cell can cover an area several miles in diameter). However, Cell ID offers some distinct advantages. In particular, it can quickly identify the approximate location of a MS (within three seconds), and it does not require any equipment or network upgrades. Because of these advantages, several coarse-grained location applications can be enabled. FindMe[5] is one such application that allows a user to associate a description (e.g., home, office) and additional attributes (e.g., ZIP code) with a Cell ID and publish the information to FaceBook.[6]

In addition to Cell ID, several mechanisms have been developed for computing the location of a mobile device by relying on the MS signal received by two or more BSs. Among the most popular of these mechanisms are *Time of Arrival* (TOA), *Received Signal Strength* (RSS), *Time Difference of Arrival* (TDOA), *Angle of Arrival*

[5]See http://electricpocket.com/findme/ for more information.

[6]See http://www.facebook.com for more information.

(AOA), Enhanced Observed Time Difference (E-OTD), and *Advanced Forward Link Trilateration* (A-FLT).

The TOA method is a network-based positioning method that uses the absolute time of arrival of the signal from a MS at a BS in order to compute the distance of the MS from the BS. Since the signal travels at the speed of light, this computation is quite efficient. In a two-dimensional environment, the computed distance places the MS at a circle whose center is the BS. By using a second BS, the intersection of the two circles provides two possible locations where the MS may be at. Using a third BS allows for the computation of the MS device using trilateration, as shown in Figure 3-2. We should note that in order for TOA to work correctly BSs must have theirs clocks synchronized. This is typically achieved using GPS.

The RSS method works similarly to the TOA method in that it uses trilateration for computing the location of a MS. Here, instead of using the signal arrival time, the propagation path losses from the MS to the BSs are measured in order to compute the distance of the MS from the BSs.

The TDOA method is a well-known multilateration (also known as hyperbolic positioning) technique that computes the location of a MS by measuring the

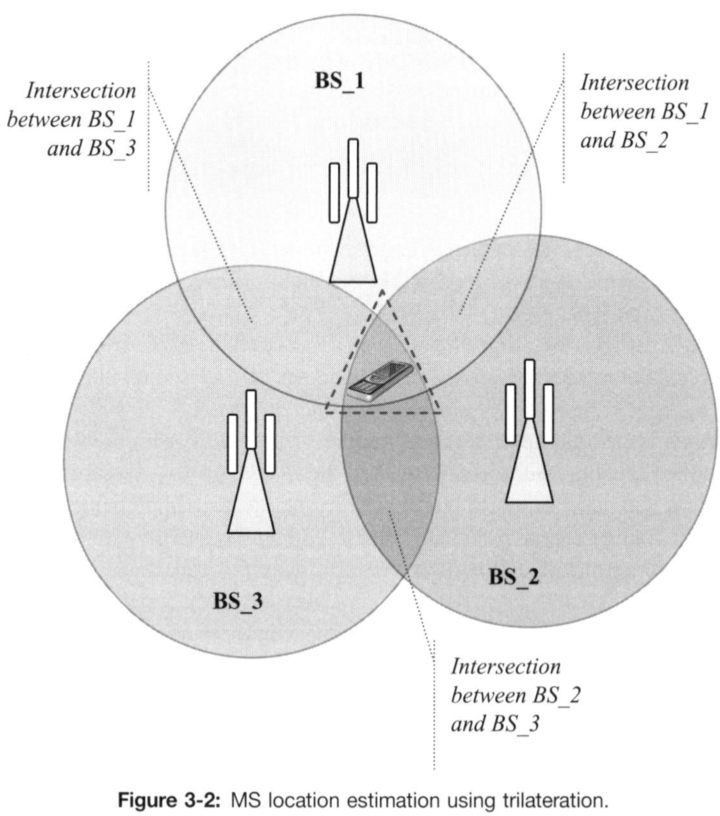

Figure 3-2: MS location estimation using trilateration.

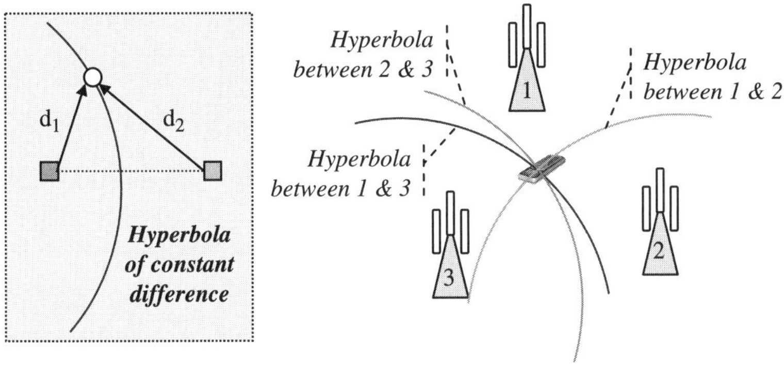

Figure 3-3: MS location estimation using multilateration.

differences in arrival times of a signal emitted by the MS at three or more well-known BSs. By computing the TDOA of the MS signal from two BSs, a hyperbola on which the MS lies is computed. The intersection of the three hyperbolas computed by three pairs of TDOA measurements is the estimated location of the MS. Figure 3-3 illustrates the TDOA method. Similar to TOA, the BSs must be time synchronized and know the distance between them. This can be achieved using GPS.

The AOA method is yet another network-based positioning method that can be used by BSs with multiple antennas. Here, the multiple antennas at a BS can determine the direction of a signal coming from a MS that is within line-of-sight from the BS. A second BS can perform the same computations and then compare the results with those of the first BS for estimating the location of the MS. This method is subject to errors arising from multipath interference (e.g., MS signals bouncing off solid objects) and requires careful alignment of BS antennas.

The E-OTD method uses software on the MS to assist in determining its location. The MS software measures the difference in the time it takes for data from surrounding BSs to reach the MS. Based on these measurements and knowledge of the exact locations of the BSs, the relative position of the MS with respect to the BSs can be computed either by the device or a network-based service. The accuracy of this method is about 125 m.

The A-FLT method works similarly to the TDOA method for CDMA networks. Since CDMA networks are inherently synchronous in their operation, the phase delay between signals send to a pair of BSs can be measured and compared against the same measurements for another pair of BSs. By using three pairs of such measurements, the approximate location of the MS can be computed.

3.3.2 GPS and Assisted GPS

The Global Positioning System (GPS) (http://www.gps.gov) is a radio-navigation system built by the U.S. Department of Defense to provide reliable positioning, navigation, and timing services. The GPS consists of three components: satellites orbiting the Earth, ground stations on Earth, and GPS receivers. There are 24 main satellites orbiting the Earth. A number of "spare" satellites are also orbiting the Earth and can by ordered to take the place of malfunctioning satellites. Ground stations are responsible for monitoring and controlling the satellites. GPS receivers receive the broadcast signals sent by satellites, process them, and compute a three-dimensional location (latitude, longitude, and altitude) and time.

Computing the exact location of a GPS receiver consists mainly of measuring the distance of the receiver from at least three satellites by computing the time it takes for satellite signals to reach it. Once this is done, triangulation is used for computing the exact location of the receiver, as shown in Figure 3-4.

By measuring the distance of a GPS receiver from a satellite, we can define a sphere centered at the satellite and having the receiver on its surface. By measuring the distance from a second satellite, the GPS receiver is positioned in the intersection of the two spheres. Measuring the distance from a third satellite reduces the possible locations of the GPS receiver to just two points. Determining which of the two points to use can be achieved by either measuring the distance of the GPS receiver from a fourth satellite or rejecting one of the points because it does not correspond to a valid location on Earth.

The accuracy of the location computed by a typical GPS receiver is between 10 m and 40 m when there is a clear view of the sky. To increase location accuracy to within a few meters for moving objects and centimeters for stationary objects, Differential GPS (DGPS) was developed. DGPS utilizes fixed reference stations on Earth that assist regular GPS receivers in computing their distance from satellites by providing them with error correction information regarding the timing of various satellite signals. These reference stations know their exact location and can compute the difference

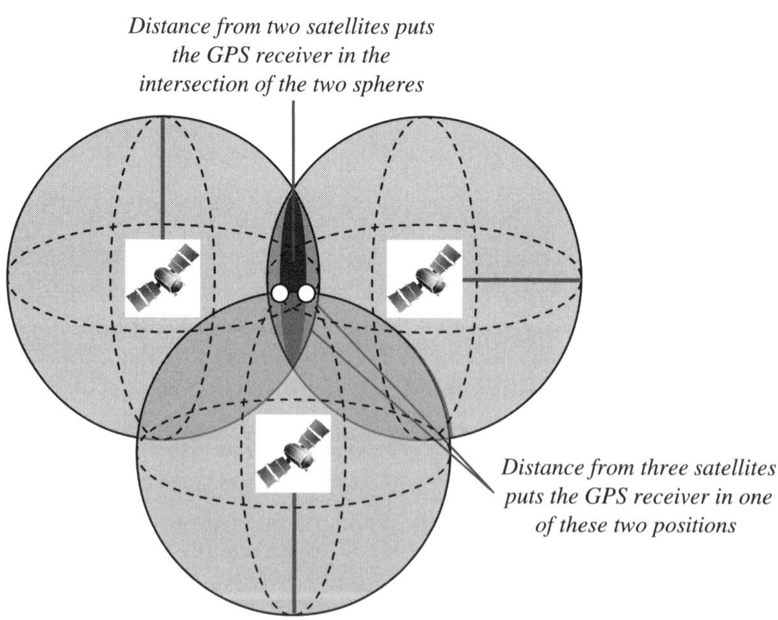

Figure 3-4: GPS positioning.

between the time a signal is received from a satellite and the time this signal should have been received in the absence of any errors or interference. Existing Personal Navigation Devices (PND) such as those offered by Garmin and TomTom incorporate DGPS for offering accurate positioning information.

Two well-known problems of conventional GPS receivers are their inaccuracy under poor signal conditions (e.g., indoors, under dense foliage, surrounded by tall buildings) and their inability to compute their position when turned on under the above conditions unless they are able to receive clear signals continuously for about a minute. These problems are addressed by the Assisted GPS (A-GPS) technology. A-GPS utilizes an Assistant Server that helps a GPS receiver determine its location faster and with less computation overhead for the receiver. The Assistant Server receives GPS satellite signals on its own and provides satellite information to GPS receivers to speed up their location computations. In addition, the Assistant Server can receive unprocessed GPS satellite signals from a GPS receiver, process them using the satellite signals it receives, and then perform location calculations on behalf of the GPS receiver. A-GPS offers an accuracy of about 30 m and allows a device to compute its current location is as low as 30 seconds.

3.3.3 Wi-Fi Positioning

The number of IEEE 802.11 (Wi-Fi) capable devices continues to grow at a fast rate. According to ABI research, the number of Wi-Fi chipset shipments is projected to exceed one billion per year by 2012.[7] The majority of these chipsets will end up in mobile handsets, with consumer electronics products and laptops capturing the second the third places, respectively. These Wi-Fi projections suggest that offering location-based services based on Wi-Fi technology has enormous potential.

Wi-Fi Positioning System (WPS) is a term coined by Skyhook Wireless[8] to describe its Wi-Fi based location system. The main idea behind WPS is the following. First, build a database that contains information about as many Wi-Fi hot spots as possible (using unique IDs and MAC addresses) by listening to IEEE 802.11 signals. Then, augment the database with information about coverage (e.g., signal strength) of these Wi-Fi hot spots at as many locations as possible (Skyhook has been doing this by having employees driving around major U.S., European, and Asian cities and recording this information using GPS receivers). Next, keep the database up-to-date by periodically updating this information.

A Wi-Fi enabled device running a WPS client can then determine its location by retrieving information from the Wi-Fi hot spot database about the Wi-Fi hot spots it can "hear" at its current location. Using this information, it can compute its location by comparing the signal strengths of these Wi-Fi hot spots against signal strengths recorded in the database for different locations. Obviously, the more densely populated the area is with Wi-Fi signals, the better the computed location accuracy will be. The typical accuracy of this Wi-Fi positioning technology is between 20 m and 40 m. WPS is available on the Apple iPhone and iPod Touch.

[7] See http://www.abiresearch.com/abiprdisplay.jsp?pressid=809 for details.

[8] See http://www.skyhookwireless.com/ for more information.

3.3.4 Short-Range Wireless Communications

A wide variety of consumer and business mobile and handheld devices (e.g., cellular phones, cameras, laptops, printers) offer built-in short range communication capabilities by utilizing one or, in some cases, more than one short-range wireless communication technologies. Short-range wireless communications technologies allow data exchange between devices that are in close physical proximity, with or without line of sight. The more commonly used short-range communications technologies are:

- Infrared
- Radio Frequency Identification (RFID)
- Bluetooth

Given the widespread use of Infrared, RFID, and Bluetooth, utilizing these technologies for computing the position of a mobile device offers several advantages over alternatives technologies (e.g., Cellular, GPS, and Wi-Fi), including better accuracy, lower costs, and increased security. In the remainder of this section we will provide a brief overview of each of these technologies and discuss how they could be used for computing the location of a mobile device.

Infrared The unregulated nature of infrared radiation, its similarities with visible light, and the low cost of infrared transceivers make infrared communications a natural choice for short-range wireless communications (the ubiquitous TV remote control attests to that). Infrared communications take place between a *transmitter* and a *receiver*. Transmitters and receivers belong to two main categories: *directed* and *nondirected* or *diffused*. A directed transmitter has a narrow-beam radiation pattern, while a nondirected transmitter has an almost omnidirectional radiation pattern. Similarly, a directed received has a narrow field of view, while a nondirected receiver has a wide field of view.

Directed infrared communications support one-to-one data exchanges between a transmitter and a receiver that are in the line-of-sight, that is, they are across from each other. Nondirected infrared communications support many-to-many data exchanges between transmitters and receivers without requiring a direct line-of-sight. Hybrid approaches that combine directed transmitters with nondirected receivers and nondirected transmitters with directed receivers are also possible [BAR].

Infrared communications protocols are being standardized by the Infrared Data Association (IrDA) nonprofit organization, which was founded in 1993. All IrDA standards use baseband modulation techniques that support low power consumption. Infrared data communications are specified in the IrDA Data family of specifications. The IrDA Data family includes the following directed specifications:

- Serial IR (SIR): This is the very first standard that was intended to replace a serial cable communication with a wireless IR connection and support a bandwidth of 125 kbps.

- Control IR (CIR): This standard targets input devices used by humans, such as mice, keyboards, joysticks, etc. It enables a wider connectivity angle and larger distance between the transmitter and the receiver.
- Fast IR (FIR): This standard is backwards compatible with SIR and supports up to 4 Mbps communication bandwidth.
- VFIR: This is one of the latest standards that enables up to 16 Mbps communication bandwidth.

IrDA Data specifications do not provide data encryption. This is because their directed communications requirements (line-of-sight), limited transmission range (less than 1 meter for most of them), and inability to penetrate walls or other solid objects (a property of natural light) make it very difficult to intercept such communications.

The position of a mobile device equipped with an IrDA transmitter can be determined by retrieving the exact position of a line-of-sight IrDA receiver by sending a "location" command to the receiver. Here, we assume that the receiver knows its exact location. While this approach can work fine for relatively small geographic areas (e.g., university campus), deploying in a large scale is challenging due to the large number of receivers that would have to be deployed.

RFID Radio Frequency Identification (RFID) is a short-range communication technology that can be used for uniquely identifying an entity, such as products, humans, animals, etc. that is static or moving, even at very high speeds. An RFID system consists of two main components: a *transponder* or *tag* and a *reader*. RFID tags are small integrated circuits with an antenna that can be attached to any physical object. RFID readers interact with RFID tags and retrieve information stored in them.

There are three main categories of RFID tags available: *active*, *semi-passive*, and *passive*. Active and semi-passive RFID tags contain an internal power source (batteries), while passive RFID tags do not. Active RFID tags may initiate communication with readers and have a range of over 100 meters. Semi-passive RFID tags do not initiate communications and have a range similar to that of active RFID tags. Passive RFID tags rely on the electrical current induced on their antennas by an incoming radio frequency signal from the antenna of a reader in order to transmit information stored in them to the reader. As such, most passive RFID tags have a relatively short range (from a few inches to several feet) and a very long life span.

RFID tags store information in nonvolatile read-only memory (EEPROM), random-access memory (RAM), write-once read many (WORM), or a combination of them. EEPROM is the cheaper form of RFID tag storage and does not require a power supply. On the other hand, RAM requires a power source and, hence, it is only available on active and semi-passive RFID tags. The amount of information stored in an RFID tag can range from a few bytes to over 2 kilobytes.

RFID readers can be quite complex depending on the application associated with them. In general, an RFID reader sends a predefined signal to an RFID tag, and when the tag responds, the reader reads the signal, decodes the data and either stores the data locally or passes the data to an application. When multiple RFID tags are within a reader's range, the reader may receive data from several of these tags. In this case, the reader may need to remember past data and compare them against received data in order to determine whether to ignore or not the newly received data. Alternatively, the reader may issue queries to specific tags directly.

Since RFID technology does not depend on line-of-sight, security is of great concern for applications that require accurate and trusted communications. There are two main security concerns, RFID tag integrity and trusted communications between tags and readers. RFID tags that use EEPROM or WORM storage are less vulnerable to attacks that attempt to alter the information stored in them. However, RAM-based tags can be altered and, hence, convey the wrong information to a reader. Current developments in the area of RFID security focus on cryptographic techniques for addressing data integrity and authorized access.

Since RFID tags can store a reasonable amount of data and transmit the data to distances of several feet, one could utilize RFID tags for providing a positioning system by deploying fixed tags storing their exact location. Mobile devices equipped with RFID readers could then retrieve the location information stored in these tags and compute their location relative to them. Advanced applications could also maintain a database containing detailed information about the vicinity of an RFID tag and access such information on demand using a wireless connection using a unique ID associated with a tag as the key to retrieving such information. This approach is discussed in [CHO] in the context of a telematics application.

From a technical standpoint, the above solution is quite feasible and, as demonstrated in [CHO], works relatively well. However, large-scale implementations of this solution can be quite challenging. This is due to the potentially large number of RFID tags that have to be installed and configured to store their exact location. Obviously, computing the exact location of the installed tags requires a different positioning technique to be used, such as GPS. Therefore, the most obvious application of such a solution is to serve as a complementary positioning technology to one of the other available positioning technologies.

Bluetooth Bluetooth is a short-range wireless communication standard that uses radio transmission in the unlicensed 2.4 GHz Industrial, Scientific, and Medical (ISM) radio band. Because the ISM band is quite crowded (cordless phones, microwave ovens, etc.), Bluetooth relies on a pseudorandom frequency hopping approach to avoid collisions with radio signals from other sources. The effective range of Bluetooth communications depends on the device power class and can range from 100 m for Class 1, to 10 m for Class 2 (the most common class), to 10 cm for Class 3. Bluetooth can support both asynchronous data communications and synchronous voice communications with an aggregate bandwidth of 1 Mbps for version 1.2 and up to 3 Mbps for version 2.0+EDR (Enhanced Data Rate).

Bluetooth-enabled devices can communicate with each other using a master-slave model. The roles of master and slave are not fixed, and a device can alternate between them. Typically, the device that initiates communication is the master. At any point in time, one master and up to seven slave devices may communicate with each other forming an ad-hoc communications network that is referred to as *Bluetooth Personal Area Network* (PAN) or *piconet*. Multiple piconet ad-hoc networks can be combined to form a larger ad-hoc network that is referred to as *scatternet*. A given Bluetooth device may be the master in one piconet and a slave in one or more other piconet networks. Figure 3-5 illustrates a scatternet network.

A positioning system could be implemented using Bluetooth piconet ad-hoc networks in several different ways. For example, a Bluetooth device may offer a positioning service that can be discovered by near-by devices using the Service Discovery Protocol (SDP) and the Service Discovery Application profile (SDAP). Alternatively, a network database containing location information associated with some Bluetooth devices could be used for looking up the positions of nearby discovered devices and calculating the position of the inquiring device. A detailed discussion about these approaches and the feasibility of these approaches in the context of fast moving devices is important in the large [HAL] [MUR].

In either of the above alternatives, additional calculations are required for estimating the position of the inquiring device. These calculations compute the distance of the inquiring device from the position associated with nearby devices by using the Effective Radiated Power (ERP) (e.g., power required to maintain a connection) or Received Signal Strength Indicators (RSSI).

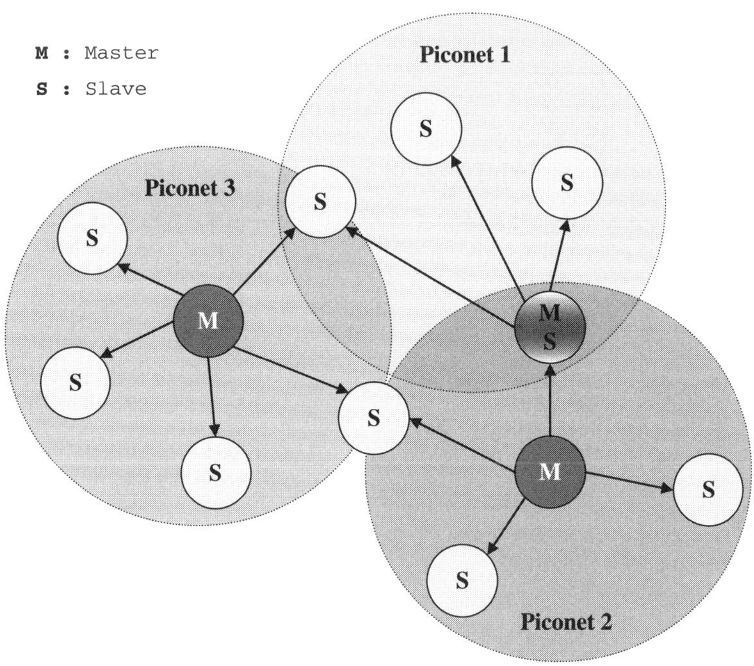

Figure 3-5: Bluetooth scattenet configuration.

3.3.5 Evaluating Positioning Technologies

In the previous sections we described several positioning technologies. The obvious question one can ask is which of these technologies should be used for enabling location-aware services in a mobile environment. The answer to this question is not always simple. Depending on the infrastructure in place, device capabilities, location accuracy, and service features, the selection may vary. We discuss some of the key evaluation criteria one has to consider before selecting the positioning technology or technologies to use in the remainder of this section.

- *Accuracy*: Location accuracy varies depending on both the positioning technology used and environmental conditions (e.g., indoors, signal quality, etc.), ranging from just meters for GPS to several miles for Cell-ID. While GPS-based accuracy is required for applications that offer navigation and turn-by-turn directions, it may not be needed for services that can handle lower levels of accuracy. Therefore, before selecting a positioning technology the accuracy requirements of the applications to be deployed on top of the selected technology should be examined in depth.
- *Latency*: Being able to determine the location of a mobile device fast is crucial for both improved usability and responsiveness of applications that depend on this. Cell-ID and Wi-Fi solutions can determine a location within several seconds, while GPS may need up to several minutes to receive the necessary information from GPS satellites. If GPS is required for increased accuracy, then a network-assisted GPS solution should be considered, such as A-GPS. When selecting a network-assisted GPS solution one has to consider whether the solution is available for both the target devices and the network on which these devices will be deployed.
- *Availability*: A positioning technology that is not available in the environment where the location-aware application will be used is obviously useless. GPS-based methods have a strong advantage because they offer global coverage and can work with any mobile device that has a GPS receiver. However, GPS-methods are not suited for indoors or other environments where satellite signals are weak. Wi-Fi methods could be used in such environments. However, their functionality depends on a being able to access a database containing Wi-Fi access point locations. The unavailability of such a database would render the service inoperable.
- *Device Constraints*: Depending on the physical limitations of some devices (e.g., CPU speed, connectivity options, battery life), some positioning technologies may have a higher impact than others. For example, A-GPS requires message exchanges with a network device. These exchanges consume CPU, battery, and bandwidth resources. Therefore, before selecting which positioning technology to use one has to examine the impact of this technology on the target devices very carefully because a heavy burden on devices will most likely result in unhappy users.

- *Device Support*: A-GPS was adopted by CDMA carriers in the United States to support the FCC's E911 mandate. Because of this, GPS chips have been added to the core chips of device manufactures (e.g., Qualcomm), resulting in the availability of innovative location-aware services in CDMA networks. GSM/WCDMA networks have been slow in adopting A-GPS, with forecasts of 2009–2010 as the year of broader device support. Wi-Fi has relatively limited support due to both its relative short life span so far and the technical challenges it faces with respect to battery drain and antenna placement.
- *API Support*: The availability of location-based APIs for developing location-aware applications is an important factor in determining the positioning technology, device, and network. While Java-based (Java ME) and BREW-based (IPOSDET) APIs exist for several devices, operator support for opening up these APIs to application developers has been challenging. In some cases operators impose certification fees and require a strict process to be followed for making the application available on the device, such as placing it on an operator-controlled Web server and then using an HTTP connection to download it to the device.
- *Hybrid Options*: Support for multiple positioning technologies may be required in some cases. If the GPS satellite signal is weak, Wi-Fi or Cell-ID may be used instead. This is approach followed by the Apple 3G iPhone[9]. A-GPS is the main positioning technology employed by the 3G iPhone. However, when the satellite signal is weak, nearby Wi-Fi hot spots are tried next. If the phone is not within the range of a Wi-Fi hot spot, the Cell ID is used as the fall back mechanism.

3.4 LBS ARCHITECTURES AND MIDDLEWARE

Location-based services have long existed in research laboratories as experimental systems and, as we have seen from previous sections, are now being offered as commercial products and services by wireless service providers. This section focuses on the typical software and hardware components one would find in the infrastructure of a business entity (e.g., wireless operator, large enterprise) that offers such services. The goal is to provide a broad view of these systems and components with respect to LBSs, but we acknowledge that in generalizing these architectures we are sure to miss some interesting and specific approaches used in isolation. The reader should use this section as a spring-board into these topics and should follow up accordingly.

At the time of writing, there are many LBSs available on different operator networks. Usually, these services have a device-resident component which the user downloads and installs. For instance, the BREW platform touts many such applications for BREW-enabled handsets and also provides an "LBS developer zone" website as a resource for developers. That portal provides a concise and convenient

[9]See http://www.apple.com/iphone/features/gps.html for details.

definition of LBS: "(Location Based Services) encompass a range of services enabled when a mobile device is aware of its location, or when a wireless network is aware of the location of mobile devices operating on it."[10] Some of the components that support LBS simply run stand-alone on the device and access position through the device's GPS interface, while others require sophisticated backend infrastructure (it is these latter ones we focus on here).

It is widely understood that some of the key overriding concerns of location systems include the following:

- Positioning and geo-coding (forward and reverse) – "location providing";
- Mapping and geographic information systems (GIS) services, including points of interest;
- Service logic, custom code, and policies;
- Subscriber profile, preferences, and privacy;
- Overall scalability, auditing, and billing.

Put another way, in order for LBSs to run effectively, one needs to be able to position customers in some geospatial area and to resolve the locations of customers and other points of interest related to these customers; this requires geo-coding of raw latitude/longitude coordinates and GIS services. One needs to know something about which services are desired by – or useful to – the customer, and to have written service logic that executes in a scalable and reliable fashion when needed. Finally, one would need to market and charge for the service. Kupper [KUP] provides a high-level view of LBS "roles," from which we can derive some further concerns for LBS architectures (beyond the above list):

- Integration with mobile operators;
- Agreements with content providers whose data enhances the quality of end-services.

The architecture and components supporting LBSs may depend on the nature of the offering and who holds primary control. For example, an "open" or open source implementation may use different components than those of a "proprietary and collaborative" system. One that is based on the Session Initiation Protocol (SIP) and intended primarily for a specific class of mobile handset may be different from one intended for more general use across multiple roles and end-user equipment.

As an example, the Open Geospatial Consortium, Inc. (OGC) – a nonprofit, international, voluntary consensus standards organization – is leading the development of standards for geospatial and location based services. Figure 3-6 illustrates OGC's OpenLS (location service) architecture, the result of calling out and defining the "core features" of a location service, in combination with GIS and routing. Note that in this view, applications (such as so-called personal navigators) are written

[10] Source: http://www.vzwdevelopers.com

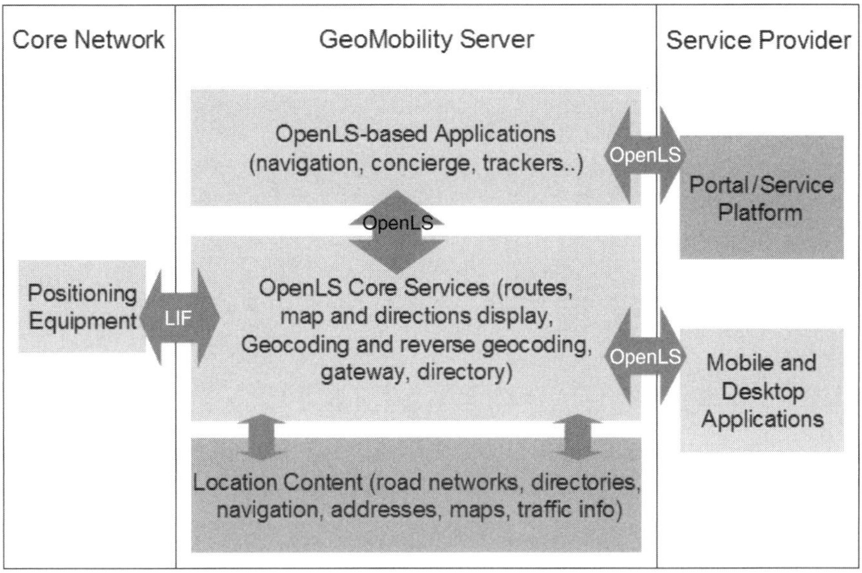

Figure 3-6: Open LS architecture [http://www.opengeospatial.org]

such that OpenLS core services (such as routing, map display, and geo-coding) are called via a well-known API. Location content (such as roads, maps, and addresses) is a separate concern. The OpenLS protocol is designed so that Service Providers can use core services via common Web protocols such as XML, HTTP, and SOAP.

The remainder of this section describes other aspects of today's middleware and focuses on individual enablers that, when used together, implement location- and context-aware systems. In this section we refer only abstractly to the underlying communication technology (e.g., IP Multimedia Subsystem networks, 2.5G, 3G, etc.).

3.4.1 GIS Databases, Mapping, and Geo-coding

A Geographic Information System (GIS) captures and manipulates geographic data. The data stored is spatial and refers in some way to objects on the Earth and is stored systematically using a storage subsystem. Additional information layers and attributes, such as those tying schools, their names, and buildings to the geographic coordinates are also a part of GIS. As all key data in a GIS database is indexed, it can be – and is – commonly used to perform analytics and problem solving. In the key areas of transportation, for example, GIS can help to provide infrastructure, fleet, and transit management. An example of a specific query to a GIS transportation system might be, "for each city in New Jersey of population greater than 25,000, tell me its distance to the nearest highway interchange." In transportation, such queries are important to state governments and highway planning departments.

A GIS system can be thought of as a layering of information maps. On the bottom of the stack is a BASEMAP, which contains only basic geographic data (such as

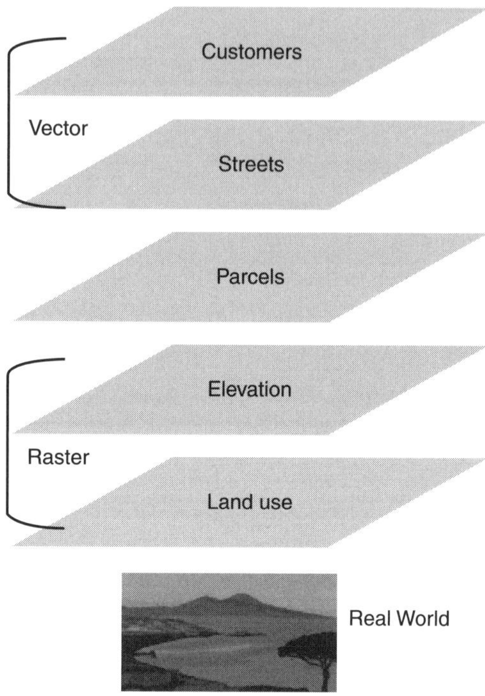

Figure 3-7: GIS Layers.

various boundaries, rivers, and streets). Layered upon BASEMAP may be a series of layers that can be leveraged or viewed in any combination. Each layer adds points, lines, or areas onto the map for the benefit of the user. Figure 3-7 illustrates these concepts.

For location-based services, a common use case is the exploitation of both user and point of interest locations (e.g., matching a user to an interesting place). To do this, the user's location is generally *coded*, depending on how the system initially understands it. If the Location Provider in the value-chain provides user A's location as latitude and longitude coordinates, then a *reverse geocoding* process occurs by which these coordinates, together with the GIS database and an appropriately formed query, are used to derive a common street address.

On the other hand, when a street address is available and latitude/longitude coordinates are needed, a *geocoding* procedure locates the address on one layer and uses other linked layers to derive the latitude and longitude. In general, a street address is found in the database by traversing street segments using a street centerline and then recursing until the street address falls within the currently active segment's address range. Storing spatial data is somewhat different from storing relational or object-oriented data. Today's approaches tend to either a) map spatial data into relational tables for a purely relational approach, b) store spatial data in external systems linked by relational or object relational databases that store attributes, or c) use an integrated approach, creating new data types to handle spatial data and store that data along with other attribute data under a single administrative control (see [RIG]). We will provide more details about this subject in Chapter 5.

Most Web users know how to make use of a very simple Web GIS toolkit supported by Google, namely Google Maps (comparable Web toolkits exist under the auspices of MSN and Yahoo). Google Maps uses the Tele Atlas GIS database of roads and geographic points of interest and computes turn by turn driving directions between almost any two points. Turn by turn, and indeed transportation data in general, is surprisingly challenging. Consider that in these computations one must store and infer using subtle information such as road dividers, turn restrictions, one-way streets, and in some cases live traffic, while at the same time delivering rapid end-to-end response time to routing queries from connected users. Other major GIS players include Intergraph, ESRI, and Navteq, while Oracle and other relational database providers offer integrated approaches to handle spatial data.

In terms of location-based services, GIS technologies are key – but invisible – building blocks. Existing networking and API technologies (such as XML schema, HTTP, and Web Services) allow remote and local GIS systems and databases to be used almost as commodities by location-based service developers. These developers need not worry about implementation issues related to spatial data. Rather, they need only worry about which queries to issue and how to process the data returned. Almost all new smartphones have the ability to run a Google Maps client and thus GIS technology is closer to most users than ever before. In the next section, we will present how one can access location information via a specific API, namely, the Location API for Java ME.

3.4.2 Location API for Java ME (JSR 179)

The optional Java ME[11] package javax.microedition.location enables developers to write location-based applications and services for resource-constrained mobile devices, such as cellular phones, using a common set of APIs. These APIs allow developers to retrieve the current location and orientation (compass direction) of the device and manage a database of known landmarks stored in the device. The APIs work with different device-based or network-based positioning methods (e.g., GPS, TOA, and E-OTD). In the remainder of the section, we will discuss the Java ME Location API capabilities by describing the main javax.microedition.location components. Table 3-1 includes a description of these capabilities.

Several of the leading wireless phone manufactures include implementations of the Java ME Location API in their devices. However, location-based applications developed using these APIs may not always work due to restrictions imposed on device functionality by mobile network providers. It is worth noting that most mobile operating systems and platforms (e.g., Symbian, Open Handset Alliance Android,[12] and

[11]Formerly known as Java 2 Platform, Micro Edition (J2ME).

[12]Available at: http://code.google.com/android/

Table 3-1: Java ME location API components

Class	Description
LocationProvider	This class represents a module that generates Location objects. The information included in generated Location objects depends on the underlying positioning technology used for determining the current location of the mobile device. When multiple positioning methods are available, the selection of which method to use can be determined by application specified criteria. The following criteria are available: – **Accuracy:** Horizontal and vertical accuracy of location coordinates; – **Response Time:** Desired maximum time allowed for retrieving location information; – **Power Consumption:** Desired power consumption for retrieving location information; – **Speed and Course:** Ability to determine speed and course; – **Address:** Ability to determine textual address information; – **Cost:** Specify whether the location method is allowed to incur costs while determining the current location. Such costs are associated with any network communications required for retrieving current location. Applications can interact with a LocationProvider object in one of the following ways: – Retrieve last known location of the mobile device; – Request retrieval of the current location of the mobile device within a specified time period; – Register proximity callbacks when the mobile device is detected to be within a given radius (in meters) from specified coordinates; – Register location callbacks by requesting periodic location retrieval operations.
Location	This class represents a set of location information. In particular, the following information is available: – **Location method:** Method used for retrieving location information, e.g., GPS, TOA, AOA, E-OTD, Cell ID, and Bluetooth; – **Address:** Textual address information (e.g., street address) associated with current device location, if available; – **Coordinates:** Device coordinates expresses as latitude, longitude, and altitude values. Horizontal and vertical accuracy of these values may also be available; – **Course:** Device course relative to true north, if available. – **Speed:** Ground speed in meters per second at the time of the measurement, if available; – **Timestamp:** The point in time when measurements were made.

(*Continued*)

TABLE 3.1 *Continued*

Class	Description
Landmark	This class represents a particular location having a user-defined name associated with it, such as Home, Office. The location is represented using latitude, longitude, and altitude values, together with vertical and horizontal accuracy values, if available. In addition, street address information may be associated with a Landmark.
LandmarkStore	This class represents a persistent storage manager for Landmark objects. Many Landmark stores may exist, each having a unique name. Within a LandmarkStore, Landmark objects can be associated with one or more categories. A category represents a thematic grouping of Landmark objects, such as restaurants.

Google Gears[13]) offer location APIs that are similar in spirit to the Java ME API. The reason for this is to foster and encourage the development of location-based services on their platforms.

3.4.3 Charging

The move from per-positioning fix charging to flat rate charging is inevitable and desirable for customers. Instead of paying every time a location service geocodes and uses her location (derived using A-GPS or some other positioning technology), the subscriber is now more eager to use location-based services free of worry of exorbitant charges. Many of the location-related services offered by today's operators – such as bread crumbing, and geo-fencing – are flat rate services. From the operator's perspective, charging may occur on the basis of one or more criteria as listed below, and may be metered (e.g., time or consumption based) or discrete (e.g., itemized or flat-rate):

- *Application level* – Charging algorithms take into account the type of information that has been delivered at a high level. For example, distinguishing charging between a ring-tone and a music video.
- *Session level* – Charging algorithms take into account the type of network functions used to support the subscribers' interactions and communications. Network functions such as voice sessions, multimedia calls, and push to talk are examples.
- *Bearer level* – Charging algorithms take connectivity and network type into account, such as IP layer and QoS issues. In this sense, the customer is charged for the medium being used, not for the services rendered or content riding on the medium.

Real-time charging refers to the customer's ability to have immediate knowledge of the cost of communication actions and to have immediate access to account balances in

[13]http://gears.google.com

order to avoid overruns. The ability to be compensated on these multiple levels is a key enabler for an operator in support of location based services.

3.4.4 Subscriber Privacy and Profile Management

Subscribers with mobile handsets are identified in the network in different ways. The International Mobile Subscriber Identity (IMSI) is an ITU-T specification that structures information about the user's handset for the purposes of signaling, but is not generally transmitted over the radio network. An MSISDN is another type of ID, stored in the wireless operator's database. MSISDN has a many-to-one relationship with a given subscriber and is essentially the telephone number of the user (users may have more than one number, thus the many-to-one relationship). As a mobile subscriber roams amongst networks, a Temporary Mobile Subscriber Identity (TMSI) is assigned when the user is first recognized in the wireless infrastructure. Since the TMSI is dynamically created, it is often safer to disseminate TMSI than IMSI or MSISDN amongst location service providers.

In general, the onus is on the network operator to safeguard subscriber identity and, in the case of location, ensure that an eavesdropper cannot compromise an individual subscriber location. In general, this is a matter of using best practices to avoid system breach and hacking. Also, the network operator may or may not have the entire or complete picture. Many of today's operators use A-GPS or network-based positioning techniques in which the positioning granularity is essentially cellular or sector region (in the order of 10s or 100s of meters). In these cases, a breach could compromise user positions up to this granularity. GPS accuracy *can* be much better – such a breach could compromise positions up to several meters. When information, such as subscriber location, is shared amongst the ecosystem of location services, things are inherently more complicated and risky as a breach could happen at any one of many interorganizational interfaces.

Grutelek et al. [GRU] clearly describe two classes of privacy risks in location-based services: those relating to location and those relating to messaging (contents of the message). They go on to identify three tangible risks: 1) "adversary" A learns that a subject S is in location L and has sent message M; 2) subject S sends an anonymous message but S's identity is derived from past compromised messages from the same location; 3) S's trajectory can be inferred if A identifies S through a message and then correlates S to previous location updates.

Recently a British insurer Norwich Union scrapped a plan for "telematics-based insurance" in which they would monitor, via GPS traces, the driving behaviors of insured customers and charge their insurance accordingly (e.g., one rate between certain hours and speeds, another rate for others). In this system, GPS data from a vehicle was to be associated with a unique key and then encrypted and sent over a wireless network to a data warehouse for further processing. The unique key method allows the vehicle to emit what, to most eavesdroppers, would be anonymous data, since they would not be able to associate the key with a driver name, license plate, or insurance policy. The private associations are available separately via secure backend databases.

Both packet and circuit-switched network infrastructures have various "registers," or databases, that play key roles in storing subscriber location and other details. Examples, using appropriate terminology, include: Visited Location Register (VLR) and Home Location Register (HLR). Such databases are key enablers for both session setup and mobility. They store locations of registered subscribers and interwork with other Radio Access Network components (such as Mobile Switching Center) in many intricate ways. To a large extent, these databases are used when a mobile subscriber needs to be paged (e.g., sent a "ring" signal). The infrastructure queries the database, determines the location of the subscriber (e.g., the cellular group, or associated MSC), and uses this information to communicate with the handset.

The HLR/VLR division is an efficiency-centric compromise that allows distribution and scalability of subscriber location information (putting it "closer" to the MSC that matters) into two tiers. When one user is called by another, the calling VLR region checks if the called party is present in its own VLR; if not, it goes up to the HLR and resolves the VLR this way. The contents of a typical HLR entry in the database include things like:

- Subscriber ID;
- Subscriber cellular numbers and profile;
- Subscriber default location (e.g., MSC);
- Services available to a subscriber;
- Other details about which infrastructure (e.g., Application servers) and triggers are relevant to this subscriber.

The so-called Home Location Register DB (Home Subscriber Service in the IP Multimedia Subsystem is similar) is an infrastructure database used to store subscriber information. In general, these databases are highly responsive but highly tuned for a given operator, making them difficult to change (e.g., schema changes).

3.5 INFORMATION-LEVEL ENABLERS

Enablers, such as the ones we just described, are at the infrastructure level. In the following we will describe two enablers at the information level: "search" and "feeds". Search and feeds are solutions to the problem of the Web-accelerated growth that makes browsing less and less efficient. Fred Wilson in his popular blog, "Musing of a VC in NYC,"[14] in a posting on "The Feederization" of the Web says: "If you look at my Google analytics data, you see that search drives only 29% of all visits to this blog and referring links drive over 40%." Fred further indicated that the Facebook news feed appeared in September 2006, initially to outrage about "too much information," but users got used to it and it is now the heart of the Facebook service.

[14]The link to Fred Wilson's blog: http://www.avc.com/a_vc/2008/09/the-feedization.html

Feeds are also the heart of the Twitter[15] and FriendFeed[16] services, as well as a host of other services. This aggregation of information into a news feed has been adopted by many of the new Web services. The company outside.in,[17] for example, provides a feed of neighborhood-related news. Neighborhood feeds have the potential of being interesting and relevant to members of the neighborhood as well as to passers-by.

Search is a proactive display of intent to immediately consume, whether that consumption is in the form of content or product. For example users are proactively seeking out information that is important to them at that very moment. "Feeds," however, are a passive way for people to keep up to speed on people/topics that interest them all the time, not just when they are seeking a specific piece of content (or a product). Consequently, feeds may, in many cases, not capture user's immediate intentions and this may also have an impact on potential advertising revenues. Consider a situation where we type "great places for hikes in central Pennsylvania" into a search engine on Sunday morning, versus a situation where we see a post about a great climb spot at midnight on Wednesday. We are more likely to click on an ad for hiking gear in the search results as opposed to the ads that appear in a news feed, but, then again, maybe not.

An important consequence of having a mobile device with us most, if not all, of the time is the ease with which information can be gathered through passive feeds and preset alerts or retrieved through searches. However, all this constant inflow of information, some of it requiring immediate attention or reaction and some of it requiring storage and indexing for future use, may create a problem of information overload. Indexing the information based on content or relevance is a hard technical problem and is very time consuming. The problem is even harder when it comes to indexing pictures and videos. Obviously, there are as many possibilities to develop and implement search and filtering algorithms as there are creative people to invent them (Google, Microsoft MSN, Yahoo and others have products that compete directly with each other, and in many cases the outperforming product is not the more well known brand).

Mobile local search, or the ability for an on-the-go user to search for information that relates to her current locale, is extremely topical. Though the technical details are beyond the scope of this text, it is worth noting that almost all major – and many minor – search engines are turning attention to getting search results right the first time for mobile users by trying to understand more of the users' context. The term "semantics-based" search is used a lot by these companies and refers to the system's ability to understand meaning at some deeper nontrivial level. PowerSet,[18] for example, has technology that exploits a deeper system understanding of both the structure of natural language queries and the information in the back in data-store. In this way, better results can be delivered to search customers. Yahoo!, Microsoft, and Google all publish research papers in which the current art of

[15]www.twitter.com
[16]www.friendfeed.com
[17]http://outside.in/
[18]http://www.powerset.com/

contextual and semantic search are used to improve search results and/or provide a better mobile search user experience.

Another interesting point to note here before we take a brief look at a few types of filtering approaches is that whoever controls the design of the filtering and search algorithms has power. The potential danger of all of this technology is that, depending on how you set the algorithm, you no longer simply search or filter reality, but actually create it. For example, a change in how the search engine works can drive business to some on-line retailers while destroying the business of others. It can literally create trends, and promote views, all through the seemingly blind eye of the program. The problem is that these programs are anything but blind, and we may be on the verge of legal battles due to the creativity of those trying to use the system to deflect traffic from the owner of a trademark, to their own competing brand, with full compliance with the providers of the search and filtering solutions.

As we show throughout the book, the mobile user context is important to better understand and anticipate needs and desires. As an example, consider a user that is driving on country road number 422 near Limerick, Pennsylvania, and is looking for a restaurant for lunch in 15 minutes. The service might conduct an automated search in which what is being looked for is not just a "restaurant" but a restaurant in or near Limerick, not too far from Route 422, which also fits the user stored profile. Conducting this geographically sensitive search is not straightforward, as keywords that specify Limerick and Route 422 may miss information items that refer to nearby restaurants whose addresses or descriptions do not use these keywords.[19] Once a set of answers comes back to the service, a filtering mechanism can pick out the relevant restaurant based on considerations that may include: types of menu the restaurants offer and which one will fit better some overall criteria, dietary information, overall calories, and so on. In forming a response the system may take into account past user actions including, for example, what the user has already eaten for breakfast that day.[20]

Now, if we follow this scenario forward, we may realize that we need to index the information associated with the restaurant. The information may include the restaurant address, the receipts for the lunch, some pictures we may have taken inside, or messages we will be sending to others about it. It may also include information about our trip (trip to Lancaster, Pennsylvania) or even who we were with. This information, if served as a set of indices, will help us retrieve the event. We can then say, "I remember the restaurant had a large porch with two white chairs in the front. We took a picture of it." And perhaps if the information about the restaurant included visual indices, you will be able to retrieve it.

The idea that physical context can be a powerful cue for recall is not a new idea. Researchers in psychology have developed theories about this sort of memory, called episodic or autobiographical memory. They observed that we naturally organize our memories for past events into episodes and that the location of the episode, who

[19]See for example www.metacarta.com for a solution

[20]See, for example, www.webdiet.com

was there, what was going on, and what happened before and after are all strong cues for recall (e.g., see [TUL]).

Various life-blogging applications have developed context aware indexing methods and organized this data into a form that mimicked the episodic memory structures that are created naturally by the user. Needing to recall a detail from a past event, and armed with our device, the user can then draw upon his or her own, possibly fading, episodic memory, to locate similar episodes and data stored in the permanent memory of the device. In this way, the user could use the small things he or she could remember about the context of the event to retrieve the details that were been forgotten. For example, the pioneering works of M. Lamming and M. Flynn [LAM] of the system "Forget_Me_Not."

Collaborative filtering technologies such as the recommendation engines used by Amazon.com, allow us to predict what information or products we may be interested in based on complex correlation of our behavior with a historic behaviors of groups of people who are like us in one aspect or another. More recently, companies like Pelago[21] allow us, in real time, to connect to our social network through our mobile device and even see their locations on a map and consult with them on subjects of importance at the moment like where to look for things. Pelago's first product, Whrrl,[22] "allows you to discover the world through the eyes of your friends and other people you trust ... take it with you into the real world on your mobile phone."

3.6 SUMMARY

As it is becoming apparent for our last discussion here, designing a mobile service is not just designing for some abstract notion of a user. What the service designer needs to look at is a user embedded in a context. This entails recognizing the different roles that people have and for which they use the mobile device, as well as the different ways they can communicate through the device – SMS, mobile Web, voice, and application. Depending on the context of the user application, expectations could be different. In the next chapter we take a deep dive into the issue of "Simplicity and User Experience."

QUESTIONS

1. Location information is very valuable and can be used to offer personalized services. However, without enforceable privacy and security policies, location information may be missused. Should location privacy and security laws and regulations be used?

2. As mobile devices start having multiple radios (e.g., cellular Wi-Fi), should location be computed at the device or the network? If at the network, how should the different networks coordinate their computations?

[21] www.pelago.com
[22] www.whrrl.com

4

SIMPLICITY AND USER EXPERIENCE

> "No one likes to suffer the frustration of waiting. Thus all of us, consumers and companies alike, often try to find ways to beat the ticking hand of time. We go out of our way to find the quickest option or any means to reduce our frustration. When any interaction with products or service providers happens quickly, we attribute this efficiency to the perceived simplicity of experience."
>
> "Simplicity is about subtracting the obvious and adding the meaningful."
>
> —John Maeda

Simplicity and user experience each have interesting histories in computing and mobility and are increasingly factors in determining the success of mobile services. The two are not always found in tandem – user experience is not *always* enriched through simplicity and, of course, many other factors may determine whether a mobile service is a boom or a bust. It is undoubtedly true that user experience and expectations are at least partial reflections of the culture of the times. Labeling a given era as one of "simplicity" versus "complexity" is tricky as anecdotal trends may signify a generation being overwhelmed by sudden change,[1] but there is evidence that today's Web and mobile culture both appreciate and demand simplicity more than ever. Indeed, "user experience" is an integral component of the "experience

[1]"Human Nature Can't Endure the Rate of Change" Source: http://blogs.nmscommunications.com/communications/2008/09/human-nature-cant-endure-the-current-rate-of-change – from-1883.html

The Fabric of Mobile Services: Software Paradigms and Business Demands, by Shoshana Loeb, Benjamin Falchuk, and Euthimios Panagos

economy," as discussed in Chapter 1. In this chapter we will provide background and anecdotal evidence of this and present some of the guiding principles and challenges of creating effective user experience and mobile services. We also examine trends and technologies that are profoundly affecting user experience and mobile services.

4.1 NEW SERVICES AND USER EXPERIENCE

Simplicity in and of itself is straightforward. Think of something devoid of complexity or complex features, having but a few parts, having freedom from artificial ornamentation, and so on, and you will have thought of something simple (e.g., an axe, a hammer, a teapot, etc.). By simplicity, we do not mean simple-minded, as that implies naivete, or some other limiting factors that make the tool ineffective. MIT Professor John Maeda describes "thoughtful reduction" as a good path towards simplicity; meaning that at each step of minimization, careful thought has gone into what has been culled and why it makes the resulting design better [MAED]. As he points out, a well-known marketing trickery is to deliver products that seem as small and simple as possible so that customers will feel that they are getting more value from them. Many iPods (though largely a good value!), with their reflectively "slimming" backplates, give evidence of this. A seeming paradox is that while new technologies enable more features, they also degrade simplicity, requiring yet newer technologies to solve complexity issues. Don Norman [NOR] points out in his writings that the complexity curve of a service through its lifecycle tends to be a "U shape," complex to start with, easing into a more simple and utilitarian form, then, as new technologies allow the integration of more features, growing more complex. Google Inc. clearly and publicly declares its 10 philosophies – heading up the list at number one is, "Focus on the user and all else will follow," which includes the wording, "The interface is clear and simple."

Why is all of this important to mobile service providers and users? Apart from early-adopters who enjoy the challenges of new complex technologies and highly adept technical users, most users seem to want service experiences that are innately easy and do not create unnecessary mental baggage. Mobile services and Web sites that do not provide this only frustrate and madden users. Common problems include users getting "lost in hyperspace" by having to navigate a great number of links and nested information (see Design Patterns in Chapter 7), or simply getting fatigued from having to perform repetitive low-information interactions. Empirical data indicates that every time customers need to click on a link in the website in order to find what they are looking for, on the average, about 25% abandon the search.[2] Users still waste loads of time in front of computers – Ben Shneiderman reports that a California-based survey found that users waste about 5.1 hours per week just trying to figure out and use their machines effectively [SHN]! Thus, over-complexity is a notorious thief of our time. Bill Buxton once compared optimal human-machine interface fluidity to the interface of a trombone [FISH]! His point was that like the trombone, we should strive to be able to interact so fluidly with services that it feels

[2]See, for example, http://www.toprankblog.com/2008/08/pay-per-conversion/ and also, www.google.com/websiteoptimizer.com for other case studies.

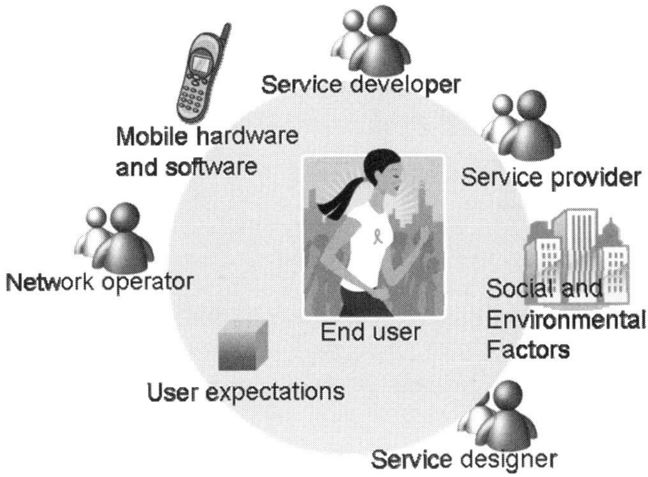

Figure 4-1: Factors and concerns from multiple disparate entities are relevant in mobile service design. Are the user's concerns at the center?

as though they are extensions to the body, perception, or will. Trombone players make visceral use of their hands, mouth, ears, even saliva to meld their senses with their instrument through its simple interface. Most of today's electronic devices, he argued, are "foreign" objects that will not allow analagous fluid natural interactions. Think of the majority of cell phones in people's hands today and one immediately sees that this is true. The clunky keyboards, the tiny squint-inducing buttons; if it were a musical instrument what sort of cringe-worthy music would it play? Things are improving though and some revolutionary progress has been made in mobile services and devices, a topic for a later section. Figure 4-1 depicts the factors and entities that affect mobile user experience.

The analysis of user experience (UX) is nowadays being recognized as more important than ever. Web and services companies now hire high level executives for a role called Chief User Experience Officer (CUEO). UX is related fairly closely with User Centered Design (UCD) and accessibility. UCD techniques – which may be employed by a UX team – revolve around designing a product or system using a methodology that puts users "at the center" of all concerns. By contrast, Accessibility Design refers to ensuring that information is readily available to users with disabilities such as visual, hearing, and speaking impairments. A selection of expectations for mobile device usability from the standpoint of the *visually* impaired is listed below (source: http://www.afb.org).

- Comprehensive voice output of information on the visual display, including
  - all menu options;
  - caller-ID;
  - indicators such as battery level, signal strength, roaming status, active/inactive call status, mute on/off, ringer profile; and
  - access to text messaging etc.

- Audio alerts, e.g., tones to indicate critical information such as
  - message indicators;
  - power on/off;
  - change in roaming status; and
  - low battery warnings.
- Displays with enhanced accessibility for individuals with some usable vision, including adjustable
  - contrast;
  - font; character size;
  - brightness;
  - freeze/slow scroll rate; and
  - zoom.

Jesse Garret has written popular literature on User Experience and uses a layered UX reference diagram to effectively illustrate all the interlocking UX issues, going from abstract to concrete:

- Strategy – user needs derivation through research and UCD.
- Scope – the main features that meet user needs.
- Structure – flows describing how the user interacts with the service/device.
- Skeleton – the navigation and/or interface design, information presentation modes.
- Surface – visual design, metaphor, graphical elements.[3]

A UX team might look at the product footprint as a whole, the interaction design, and the visual interface metaphor, design, and effectiveness. Contrast this with engineering and development teams, who are primarily involved in implementation of system functionality and system architecture. UX teams are sometimes staffed with both marketing and product management personnel who bring their expertise to bare in defining the product features and value proposition. As an example, while Nintendo's GameCube was often considered toy-like because of its form-factor, one strategy was to ensure that Wii was not *overly* toy-like and would nestle in comfortably amongst other audio-video equipment in the living room. Wii's controller also had to be *significantly* simpler than previous generations, wireless, and had to be accessible to users across generations.[4]

Measuring UX is not easy but service providers need metrics in order to evaluate whether their deployments are working well, both at the network level (e.g., end-to-end response times) and at the interaction level within the service (e.g., clicks and tasks). Wireless carriers, for example, have several means for evaluating their end users' quality of experience including using feedback forms and carrier-controlled

[3]Source: http://www.jjg.net/

[4]By most accounts Nintendo succeeded; Wii has seen take-up into demographics and uses that previous consoles failed to achieve (such as a tool for seniors to help promote activity).

simulated mobile handsets that interact with the network in large-scale simulations. Analytic engines can also derive network and service usage by analyzing reams of data "sniffed" by the carrier and stored in a database; the carrier achieves this by installing probes in various places in the network and logging data. As pointed out in [SHA], a carrier interested in improving customer quality of experience would have initiatives in some or all of the following:

- *Assessment* – an initial analysis of current issues accomplished via computational, survey, or other methods.
- *Monitoring* – an ongoing (but targeted) data-collection and analysis effort that helps further identify and prioritize issues.
- *Analysis* – rigorous analysis of collected data and study of impact on bottom lines.
- *Improvement* – ongoing analysis of the gains achieved by making customer experience changes.

So, both new services such as mobile services, and hardware such as the Wii, sometimes require the rethinking of user experience. The proliferation of short message service (SMS) (and Internet access) was one of the reasons that mobile handset manufacturers now commonly embed sliding keypads in their devices, and why a huge variety of keypad inputs continue to evolve (e.g., fabric keypads, touch keypads, dual QWERTY-numeric keypads, pushbutton, stylus based, thumb-as-stylus keypads, etc.). This rethinking, now seemingly obvious, remedied the surprisingly awkward keypad texting that was performed in previous generations (and is still today on some older phones). Similarly, new mobile services sometimes require rethinking, refining, simplifications, and the application of UX best practices. Emerging services like location-based dating, geo-fencing, child location-tracking, mobile social networking, and so on (see Chapter 3) may not require new hardware or keyboards but will require a rethinking of interface, interaction, functional design, and even billing in order to be most effective.

Figure 4-2 depicts the evolution of mobile user experience through the wireless generations. While earlier generations of networks provided basic communication capabilities such as SMS, e-mail, maps, and Web capabilities, emerging 4G and 5G networks will enable dramatically new mobile services. Characteristics will include immersiveness, multiple fragmented persona management, proactivity (e.g., virtual assistants), and will almost definitely serve health, wellness, and telematics (e.g., navigation, directions) needs. While some think the communications component of the phone may disappear completely (e.g., into the clothing or skin), display capability such as folding, paper, heads-up, or organic display will remain "visible" and tangible.

4.1.1 Hiding Complexity

An *effective* mobile service is easy to access, simple to start, requires little configuration, and has a clear affordances. The short message service (SMS) was a

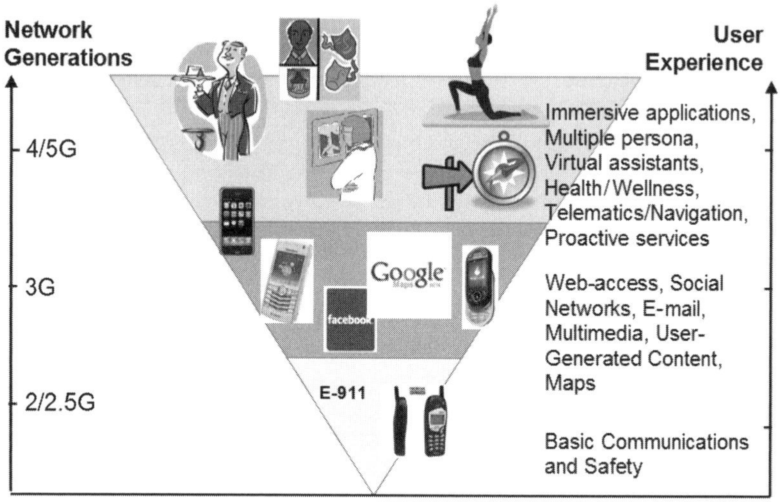

Figure 4-2: Evolution of user experience through the mobile generations.

breakthrough (somewhat an "accidental" success) service. SMS message-sending is easy to access on almost all phones (just turn it on), simple to transmit (because contacts are already stored on the device), and easy to receive. Mobile devices marketed primarily at high-frequency SMS and e-mail users incorporate features that simplify message-sending. An example is progressive contact look-up for which the user simply opens the keyboard and begins to type any name in the address book. As characters or digits are pressed a list of matching contacts is pared down on the screen (starting from the first key press); at any time the user can select one of the presented ones to complete it. Textfield auto-complete (now found on many search engine sites) does something similar with keystrokes comprising search terms. Other simple and effective mobile services include Push-to-Talk over cellular (PoC) (half-duplex voice communication), and Voice over IP. Voice over IP (VoIP) applications tend to require some set up by the end user but, once configured, the user is unaware that some voice sessions are using Internet Protocol while others may be circuit switched. In the authors' own work, an innovative approach takes the onus off of the user when her mobile handset has the option of choosing one or more access networks available at a given time and location [FAL3]. Instead of a manual network selection by the user, an agent on the handset and a centralized server collaborate in order to decide which access network is best. The user, meanwhile, simply dials a number on a special dialer and the voice session is set up on whatever access network was chosen by the agent and server (e.g., a VoIP call, cellular call, WiMAX access, etc.). Thus, the user interaction pattern remains simple and the user is not forced to understand the complexity of network characteristics.

Mobile services are highly complex and ride on complex network technology. This complexity need not – and must not – be visible to the end user as the services context and experience are not at all related to the technology that makes it happen. For example, the architecture of VoIP applications depending on the architecture, may involve many tens of systems and servers including: the SIP protocol, registries, routing algorithms, compression, billing, rating, auditing, and so on (see 3GPP white-papers and standards). Andersson et al. [AND] emphasize the complexity of today's mobile handset software and hardware and point out that they indeed *need* to be complex in order to perform up to expectations. From a software point of view the mobile device software stack can be thought of as multi-tiered, with applications running on the highest conceptual tier supported by middleware services such as network adapters, data communications, and codecs. In the middle we find sophisticated software platforms and operating systems such as Java execution environments and Qualcomm BREW. A hardware abstraction layer underlies the middleware, making the lower level hardware transparent but accessible to the higher layers. [AND] points out that even a handset circa 2005 had a whopping 6000 system-level requirements! While telecommunications services facing the mass-market customers clearly face a challenge to hide complexity, the computing world currently already finds itself in a simplification trend. Figure 4-3 depicts how complexity is hidden from mobile users in many ways.

For many years the "ubiquitous and pervasive computing" community has offered research and technologies that support invisible always-available computing. Smart-homes, for example, use such techniques to integrate into the daily lives of the inhabitants. A long-standing issue in human-computer interaction research has been making the interface intuitive and unobtrusive, and research by Mark Weiser and MIT Media Lab is exemplary. For Weiser and others, the point of the interface is not to call out the human-computer interface but to facilitate the flow of information

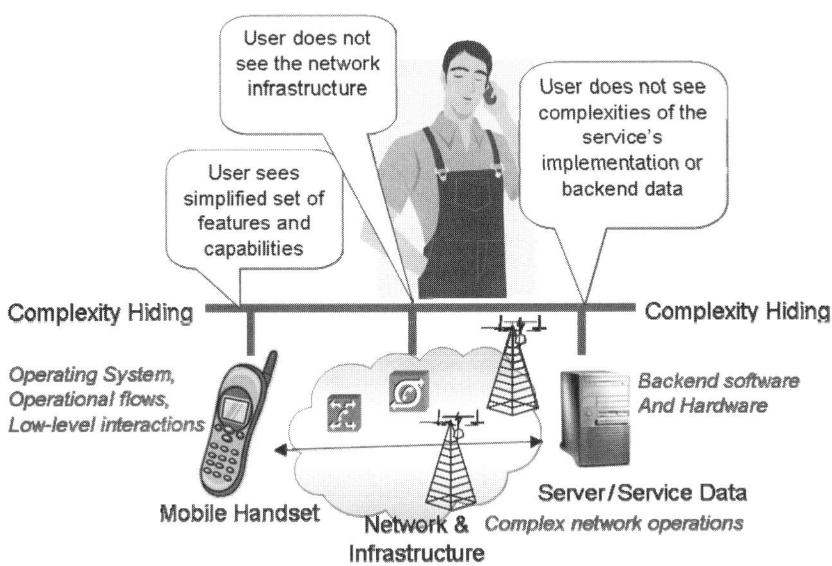

Figure 4-3: Hiding complexity on various levels.

between human and computer. Interactive art exhibits have long needed to both simplify and make invisible any mechanical and computational interfaces since these artifacts tend to detach the viewer from the experience. At New York University's Media Research Lab, Jeff Han spearheaded an innovative touch-sensitive interface-free computing paradigm. A so-called multi-touch screen awaits contact from the human hand or other pointing/touching device and handles the simultaneous input of touchpoints, as opposed to a single touchpoint. Multi-touch applications tend to be exhilarating, making us realize that when traditional limitations of the interface disappear we get more out of the application logic. Touch interfaces have the potential to be far more intuitive than mouse-based ones. Thus, a sort of simplicity is gained, freeing the user from the constraints of input devices. Touch and multi-touch handsets arrived on the U.S. market in about 2007 (Apple, Samsung), and multi-touch screens arrived with Microsoft Surface at about the same time.

There are few things as complex as trolling and indexing the Internet's vast unstructured data, and yet Google's search page interface consists of a single text field for unstructured query (some complicated query flags *can* be used). Google has led the so-called Web 2.0 trend in this regard and many other Web sites have copied their approach, sometimes even at the expense of over-simplification. For example, some sites use the simple and clean search text box approach but do not support complex expressions within the query. Sometimes (depending on the user and the search context) the user is better off presented with a more complex search screen wherein detailed search parameters may be entered. Many search engines went so far as to change the look of their main pages to mimic Google's simple clean look. In a thoughtful note, Shneiderman wonders why interfaces cannot start out simple, offering only basic features to users, and over time become gradually more complex, adapting to the users' increasing sophistication with the tool. While this vision is seen in research initiatives (e.g., adaptive user interfaces) it has not largely been seen in practice likely due to the difficultly and risks of refining the interface in a way that neither overwhelms nor annoys the user. Microsoft Office's personalized menus, in which most commonly used menu items are made visible while others are made invisible, are another very simple embodiment of the concept.

In summary, hiding complexity comes into play on many levels and domains. Mobility, cellular communications, and mobile search are all complex things – in general, end users benefit from simple effective abstractions of these complex things, unless they are specifically tasked to work with the underlying minutiae. Brad Burnham puts it into perspective nicely: "[system designers] have to seduce those users, hiding complexity, revealing one layer at time, always enticing, never intimidating, until the user one day finds they are intimately familiar with power and the pleasures of the service" [BUR].

4.1.2 Mobile User Expectations

Modern communications networks are undergoing massive changes. These changes are forcing mobile handset designers to keep up with escalating user demands and expectations (the users see more bandwidth available and want to make use of it via new services). As service providers promise and deliver high-speed networks, handsets will need new sorts of hardware architectures and new and better ways to present the multimedia information that the networks are capable of delivering: IPTV, streaming broadcast television, videos, music, video-conferencing, and so on. In 4G network scenarios in which multiple types of high-speed networks provide blanket coverage for roaming users, these customers will come to expect seamless handoffs and integrated communication sessions across networks – users will not be aware of network handoffs as they are occurring. Users will also expect continuity of services between the gamut of mobile and portable devices. A movie playing on a mobile handset will be expected to pause and continue to play on the user's home computer when he arrives home and enters the house.[5] Handsets will be multi-mode – that is, able to use two or more different types of network technologies to create communications sessions. Dual-mode handsets offering Wi-Fi and cellular service now proliferate, while multi-mode handsets supporting WiMAX or LTE access networks will deliver even higher speed networking.

Although all users expect that their handset enables high-quality voice sessions, they also want to feel like their handset reflects, in some sense, their personality. The wide variety of mobile handsets geared to different niche customer types is the device manufacturers' response to this. The ability to deliver a high-quality user experience compels network operators to have the best possible network planning and Operating Support Systems since, if calls continually drop, it will not matter if the phone has fancy features, the customer will choose another provider. As an example, the market currently bears a large number of handsets geared to customers who are heavy text message users featuring sliding full QWERTY keyboards and integration with Instant Message systems. On the other hand, the initial iPhone rollout in 2007 was, in some opinions, marred by the poor quality of AT&T's Enhanced Data GSM Environment (EDGE) network (*a Wall Street Journal* review referred to the network as "pokey"). In hindsight, the network issues have had minimal long-lasting impact on the iPhone outlook and have been addressed with the 3G iPhone.

Quantifying user experience is difficult and is a research topic of interest in both academia and industry. Carnegie Mellon researcher Niraj Tolia has studied the impact of network latency on mobile services on both thin and thick clients [TOL]. They have found that indeed *latency* is more critical than total *bandwidth* in mobile services; this is partly attributed to known work in the social sciences indicating that negative experiences are more impactful than positive ones during judgments. He notes that:

- Response times below 150 milliseconds generally do not affect user productivity.
- Response times above 1 sec create unhappy users.
- Between 150 milliseconds and 1 sec, users become increasingly aware of response time and user experience becomes tenuous.

Thus, a mobile game that stalls only a few times in an otherwise good session will still see the user come away with negative feelings towards the experience.

[5]3GPP and other standards promise to make this a reality; for voice sessions, the Voice Call Continuity group addresses session-handoff.

For a large part, mobile user expectations are set by two things:

- Marketing
- Word-of-mouth

While the perceived quality of service is actually influenced by:

- Network attributes (such as latency, bandwidth)
- Handset hardware and response time
- Application software design and usability

Product marketing and packaging is highly influential in setting the mobile user's expectations. As Sun's Michael Powers points out, a multi-player game-maker can raise users' expectations by marketing the game as a fluid real-time first-person shooter (FPS) type game (the graphics on the packaging and advertisement artwork are also vital). If the game is to be offered on a high-latency network or on devices with limited computing capability, the product team may be better off designing and marketing the offering as "Strategic," or a "thinking-man's" game, downplaying the FPS aspect and playing into the capabilities of the network. Turn-taking strategy games are much less sensitive to network latency than FPS. Word-of-mouth (buzz) marketing is also of key value. A savvy Web campaign that, for example, gets bloggers, gamers, technology Web sites, and comparison Web sites talking about the upcoming offering can raise user expectations (expectations can also be lowered by negative word of mouth). In some cases the final offering meets expectations and in others it does not. The high-profile console games Halo 3, Bully, and the Grand Theft Auto series all benefited from large amounts of expectation-raising "buzz," which in *these* cases translated into good sales. Both the Helio Ocean and Apple iPhone mobile handsets benefitted from positive pre-offering word of mouth.

4.1.3 Attention's "Long Tail"

While users have expectations as to how mobile services will integrate and improve their day-to-day lives, they are still generally hard to please. This is particularly true today as freedom of information allows users to better understand their place in the value-chain. The so-called "me" generation – i.e., roughly, those born after 1973 or so – has recently been the topic of sociologists. At times self-centered and shiftless, the "me" generation is also delighted with technology and optimistic about the world. Blink marketing [GLA1], the fast-cut sequencing of today's music videos and television, and the quick adoption and abandonment of norms and technologies (e.g., MySpace, Facebook) reflect this generation's values. When using mobile services, these users are characterized by:

- Unwillingness to wait long times for information download;
- Preference for intuitive instruction-free applications to those that require practice or training (e.g., see Nintendo's Wii device, launched in 2006).

One might also think of attention spans as being a power distribution, not unlike the "long tail" described by Anderson, in which he argues that vendors can make good revenues by selling small volumes of less-popular items to a large number of people [AND1]. The "long tail" of attention might describe the distribution attention spans through the population. Previous generations' attention spans can be thought of as sitting near the start of the power distribution, most users on average will have moderate attention spans (middle of the curve), and today's "me" comprises the long tail of users whose very short attention spans are difficult to market to but with the right approach can be marketed to both effectively and lucratively. For example, an application designed for today's elderly should focus on clear usability rather than quick startup and attention-seeking visual gimmicks [CAR], whereas one for a younger generation might enable the rapid context-shifting within the application, visual gimmicks, and other attention-attracting techniques.

In both theory and practice, attention economics and attention modeling are real research areas studied by those interested in many aspects of computing, including mobile application development. Attention economics often refers to Michael Goldhaber's thesis that, "the ethos of the old economy which makes it often bad taste or a poor strategy to consciously seek attention seems to be giving way to an attitude that makes having a lot of attention rather admirable and seeking it not at all to be frowned upon." If you are after attention, then the Web and the Internet and mobile handsets and all other sorts of information delivery mechanisms are tools of the trade. Eric Horvitz from Microsoft Research has studied the so-called *attentional user interface*. In his work, systems are enabled to observe attentional cues from humans and also to anticipate future attention through perceptual sensors (e.g., gaze tracking technology) and so on. The goal is to deliver information to the user that is attention-sensitive; that is, not only would the system deliver notifications of new emails but also would derive the urgency level of the email and evaluate the expected cost of delayed review of the email. Such work moves us viably towards mobile services that can gauge every information bundle's importance to the user and, by deriving the cost of many possible series of actions, can suggest to the mobile user an optimal series of next-steps in order to best continue their day. The intelligent computerized assistant, as envisaged, for example, by Apple's 1987 Knowledge Navigator video,[6] captured many of the notions of the attention economy and attentional interfaces. The intelligent computer assistant "jumps into" a conversation between two human speakers when one speaker forgets a fact that the agent lookups and vocalizes.

4.2 USER-CENTERED SIMPLICITY AND EXPERIENCE

In creating mobile services for the end user nothing is more important than the user experience. A high level of user experience is usually the result of a well-conceived user-centered design (UCD) strategy. The UCD process sometimes proceeds roughly as follows: 1) requirements gathering (user studies, interviews,

[6]http://video.google.com/videoplay?docid=-5144094928842683632

etc.); 2) requirements specification; 3) design (including prototypes, participatory design, etc.); and 4) evaluation (questionnaires, in situ studies, and so on). UCD, then, is the process by which the designer – concerned with the usability and usefulness of the designed product – applies guidelines and methods in order to try to meet the requirements of users. The UCD process involves early and significant face-to-face interactions with the user-base (or representative users). With regard to services, UCD is applied carefully to everything from mobile device operating systems, to Web sites, to physical devices, and mobile services. Nielson, Shneiderman, Norman, and others have written fairly extensively on this topic since the 1980s.

Mobile user experience, on the other hand, is the set of sometimes intangible or "soft" metrics relating to the effectiveness of the service both on an end-to-end and local basis. Instead of focusing solely on transmission issues (such as throughput, delay, jitter, and loss) as one would do if network-level Quality of Service (QoS) were the only item of interest, user experience is additionally gauged with qualitative measures. The following are some key evaluation questions that the user might ask herself in deriving a user experience rating:

- Suitability – Is this service delivering information that is suitable to my current request or my current situation?
- Cost – Is the service appropriately priced?
- Personalization – Is the service delivering any sort of customized information to reflect the known aspects of my preferences and context (e.g., location).
- Lifecycle – Do I have the ability to easily modify, opt-out, or quit this service?
- Delivery and security – Is the service delivered in reasonable time over the network and without compromising any information that I deem private?

4.2.1 Illustrating User-Centered Design Decisions and Principles

With the rise of mobile services that in some cases have the sophistication of Web sites and rich services, UCD remains as important as ever. Let us go briefly through some of the key steps in UCD with respect to a fictional mobile service called *Local Insure* (or LI for short) intended to allow mobile users to browse car insurance quotes on their mobile handsets offered by local insurance offices near their current location. If an offer looks good the user has the option of connecting to the service through text messages or a phone call, or, the user can be given driving directions to the office and a contact name.

First, the designers perform user-need analysis by talking with prototypical users. They would ask questions like, "What is most important to know about the insurer when getting a quote?" and "What kind of interaction are you able to perform when you have limited attention?" and so on. After some time the designers come to understand the physical and attentional limitations of mobile users (who may be driving or walking); one possible ramification is a decision to design a very bold and clear way to present quote information on the small screen of a mobile handset. The designers also come to understand what it is that users need to accomplish with the service in order for them to consider it worthwhile; one ramification of this may be to eliminate use cases that were previously thought useful by the designers and focus on the ones identified by users. Perhaps, for example, the designers originally envisaged a screen that presents comparisons between quotes but the users have indicated that such detail would not be useful on the small screen and would be difficult to read anyways. Thus, the designers could adapt the vision to present a greatly simplified quote-comparison screen.

Typically, the analysis of *what users want to accomplish* results in a set of documented user behavior patterns and a breakdown of the system into a series of core tasks. Using one of several available technologies, a mockup system might be put in place by the designer. The mockup would run on the target device and operate almost "like real." A series of user studies comprising usability testing follows. If users have problems running the service and completing the tasks that are important to them the mockup phase is the best time to discover this. For example, it may be that in the LI service the designers implemented a feature in the mockup to let the user estimate and enter her current insurance premium. During trial runs users may indicate that doing so is onerous and the team might then decide to omit the dialog completely in favor of using profiled information based upon the user's current ZIP code instead.

In reality, UCD is both difficult and important, especially given the wide range of mobile device configurations. Norman's writings [NOR] are valuable and call out several core principles. In the next few paragraphs we relate a selective and representative set of those pearls of wisdom and use the Local Insurance quote service as a running example. Note, however, that the principles have a general applicability.

A System Model Should Bridge the Mental Models and Expectations of the Service Designer and the User Norman argues for systems for which the user does not need to read a user manual. Rather, after a bit of direct interaction the user has a "*Yes, I see*" moment and is able to make progress. They can do this either because the service operates consistently with the users' mental model of operation or the user has understood the model of the service offered and mapped it to their own understanding. A principle challenge here is that the system designer and the user are different people, each with different social and cultural backgrounds. This is the reason that service testing and surveys are useful early on in the service lifecycle. As an example, imagine that for the designer of the LI service it was key that the service be requested to do something, for example, that the user had to press an "activate" button each time she wanted the system to attempt to present a local insurance quote. Imagine now that the user sees the service as a continuous "concierge" and assumes that the service will stream insurance quote information to the mobile handset intermittently and without prompting. This is a case of conflicting mental models of the service and will reduce service experience if not remedied (we mean on a different level than functional system requirements).

Simplify Tasks Task simplification is one approach for UCD. Here we see menus and options being reduced to a small number of items and we see service designers taking into account the average user's ability to keep multiple contexts in mind at once. Norman suggests that five unrelated items is the most that users can and should be asked to "remember" during service operation. In the LI service a key design decision might be to simplify the user's possible actions down to a few clear input types that require minimal overhead be remembered by the user but result in valuable information. For example, the search for insurance quotes could be kicked off by a single large button on the interface; results should be presented in a single compact screen and should not force users to navigate a number of connected forms or screens.

Don't Take Away Control Service designers carefully weigh the options of control. While it does vary from realm to realm, users want services to do great things for them but also want to retain some aspects of control. Norman points out that removing the user's ability to control the service makes the user feel like a passive information slave to it. Most people have bosses at work and nobody wants their mobile services to feel like another. In the LI service the designers need to carefully weigh the tradeoffs between providing some configurable and interact features versus providing the insurance quotes seamlessly and automatically on a constant basis. With some control the users may feel some ownership of the information delivered by LI instead of "tuning it out."

In summary, User Centered Design is a key part of the mobile service lifecycle. The rest of this section describes how the diversity of different aspects of mobile devices and services is important and how best practices and standards have emerged to bridge gaps.

4.2.2 Complexity Issues Facing Both Users and Designers

Millions and millions of mobile handsets become obsolete and are discarded each year. We are all familiar with images of heaps of used cell phones in a field somewhere as a symbol of one of the negative impacts of technology on our environment. Some estimate that the "relevancy span" of a phone is about 18 months; though it may actually be functionally operational for much longer the user is compelled to discard it after this timeframe. Marketing forces, peer pressure, and changing technologies all figure in to why this is so. For the mobile service designer it is essential to choose well the service platform. When considering a particular candidate platform, factors that influence the designers include:

- Cost – What are the costs associated with developing for this platform?
- Availability and demographics – What is the penetration of this platform and what are the demographics of the users?
- Life-span – What is the expected future of this platform or technology?
- Openness and ease – Is the platform easy to develop upon? Is it open-source? Do the devices support the necessary input, processing and storage options?

Designers are not the only ones interested in battery capabilities and life-spans. An inescapable and mostly unwanted task facing almost all mobile device owners is the management of the device's battery and its charge. Today's wireless devices rely largely on nickel metal hydride and lithium-ion battery technologies but ever-increasing power requirements from devices components (such as the color screen and the network interfaces) create a cycle of ever-shortening standby and usage times. Newer technology like lithium-polymer, zinc-air, and direct methanol fuel cell promise much longer uses, but until then the consumer must manage battery usage wisely. While the onus of managing battery life is not overbearing it does force the user to face some of the device's complexity head-on. The "perceived affordance" of the battery level bars is a simplistic and effective metaphor for the devices power constraints, but we are left wondering:

- Why does the phone not warn us if it anticipates the battery will run out during a present voice call?
- How accurate are the battery bars?
- Which modes of the device and on-board services use up most battery (e.g., Wi-Fi networking, voice session, photo-browsing, camera)?
- Why doesn't the device automatically disable nonessential aspects when the battery gets low?
- When in a voice session with another user, wouldn't it be nice if you always saw the status of the battery level of the *other person's device*?

Recent and ongoing research initiatives have indeed focused on battery-aware computing and operating systems for mobile devices. Some of these have proposed new architectures that would allow the phone to anticipate when the next charge will be (based on past profiles and activities) and if the battery is not expected to last until that time and place to then warn the user. Other work maps energy consumption to specific code components. Few real deployments of these sorts of approaches have been seen in the market.

4.2.3 Customizing and Identifying

Almost every mobile device is marketed at a particular consumer segment. At an individual level, mobile services need more and more to remember details about the *individual* user (not just the demographic) and use them to improve user experience. As a simple example, Google stores your search results and lets you review them at any time; it also presumably builds an aggregate profile of your search tendencies over time which it may use to tailor advertisements, presentation, and so on. Several technologies allow this to happen and they are reviewed in brief here.

The use of browser cookies is a well-known technique in which Web sites create and store the "state" of an otherwise "stateless" HTTP browsing session on the browser's computer. The stored file (i.e., cookie) encodes details of the user's interactions with the site and can store information between browsing sessions. Cookies are supported by all desktop Web browsers and are widely used to implement Web site-based "shopping carts," password-protected logins, and to save form data for future page visits. As a whole, the industry is realizing that personalization – especially in mobile e-commerce situation – is more and more important. Marketing is no longer a static banner ad but rather integrates data that is online and offline, happens on a continual basis as the user browses the mobile Web pages, and is better thought of as a conversational process (i.e., every good salesman engages the customer in a dialog that encourages sales!). Additionally, mobile commerce happens on a continual (if multi-session) basis through a "shopping cart," and can occur across several sites. For example, someone buying baseball bats at one store may package it with batting gloves from another. The industry is realizing that tracking and learning about mobile e-commerce customer behavior is extremely advantageous and several startups have technologies that let commerce providers track, target, and understand the complex relationship between shoppers, communities, site views, Web page transitions, and buying patterns. On the mobile devices, however, cookies are not supported consistently, and most browsers let users delete their cookies whenever they want. Sometimes, a mobile carrier may strip cookies out of the HTTP requests and responses.

Identity management technology allows users to authenticate and establish their identity with a service. Once identified by the service they may then be able to use multiple services securely and privately without the need for separate sign-ons for each. The Liberty Alliance (LA) is an example of an identity framework. LA provides a standard for single (or "simplified") sign-on (SSO) with decentralized authentication and open authorization. It allows mobile consumers to manage their personal information securely and on their own terms. In the bigger picture, let us say that a mobile user visits a variety of m-commerce sites where she would like to perform transactions. Examples include rental car sites, Amazon (books, CDs), shopping, banks, stock trading, and so on. Even though this user frequents these sites often she will have to enter her name and password onto the ubiquitous login screen at each site. With SSO technologies, the user signs-on *once* and is automatically signed on other LA-enabled sites when necessary. For example, if the user is browsing an LA-enabled airline service for tickets and needs a rental car, the switch of context to the rental service site can automatically authenticate the user, taking her immediately to the rate quote instead of to the login screen. In general this "circle of trust" includes:

- The User – (a.k.a. principal) individual or corporation involved in transactions.
- Identity provider – authenticates, stewards identity information.
- Service provider – provides goods and services to a user.
- LA-enabled clients (and proxies) – an entity that gathers the information that about an identity provider related to a user.

and all of the entities share a tacit business agreement to enable transactions in a secure and seamless fashion based on the LA standards (Fig. 4-4).

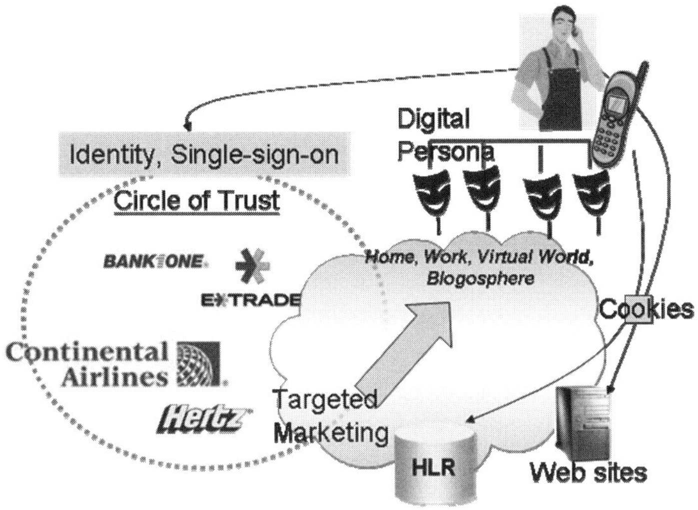

Figure 4-4: The elements at play in simplifying access to services and information in mobile communications.

Beyond sign-on to services, we note that more and more users think of themselves as having multiple digital persona. There is the "work-me," the "home-me," the "tennis player me," the "avatar-me," the "blogger-me," and so on. To make matters more interesting, each (or many) of these so-called identities may have a unique identifier, communication endpoint (e.g., e-mail address), a unique voice or attitude, and differing information requirements. If mobile service providers are to target a user, would it be better to target each of the digital persona separately or to try to target the user as an inseparable whole? How does one get access to the various communication endpoints of the user? Can the provider relay information to the user while she is playing inside a virtual world?

In mobile cellular networks, operators make use of Home Location Registers (HLR) to store pertinent information about roaming mobile subscribers. The HLR is a database server in the infrastructure (not "seen" by the end subscriber) whose main purpose is to store associated Mobile Switching Centers with which subscribers are currently associated, as well as equipment identity, and class of service. HLR can also store additional information at the discretion of the operator but always strives for rapid response times and high availability. So HLR is just one method that today's operators use to store and understand the customized information that varies from subscriber to subscriber. But while telecom operators can fairly easily create a logical mapping between a user's "home" and "work" information, current trends indicate that the number of personas for a given user will fragment into many more aspects than just home and work life. With multiple virtual worlds, blogging, gaming, social networking, and hobby persona ready to go, personal identity management and customized services are still a research challenge. For service providers hoping to market to today's user, success may be based on catering to the multiple user personas and by considering the virtual worlds they inhabit. Some mantras (adapted from [GON]) to consider include:

- Determine your company's role in providing access to knowledge, social status/reputation, and achievement or responsibility. Organize and target online products and services based on the customer's journey toward self-actualization.
- Sell to the persona, not the person. Collect psychographic data to understand online persona behavior and its interaction with others.
- Develop and retain or outsource new skills to attract, connect with, contribute to and gain insight into personas and virtual environments.
- Develop strategy, process and technology around relationships with persona bots, as a tool of mutual exploration.

Therefore, while the management of some aspects of multiple identities can be simplified by standards, other aspects are more challenging and still await solutions, such as how to target, market, and communicate with each mobile/online personality of a given subscriber to yield an effective conversation that meets the goals of the user and the service provider.

4.2.4 Gadgets and Mashups – Building the Mobile Service Yourself!

More and more, users have the ability to create their own "services" for their mobile handset. Mashup and so-called Web 2.0 technologies will be covered in a later section in this chapter; some basic Design Patterns of mashups will be presented in Chapter 7; here, we outline a few issues related to user-centered simplicity and design. A mobile mashup can be thought of as a mobile service or application that uses application program interfaces (APIs) to fuse together data from other information sources into something interesting or value-adding. Most importantly, mashups can be designed and implemented by end users themselves, a departure from the mobile landscape of not long ago when only operators and handset manufacturers had the tools to provide services. A mobile mashup might, for example, fuse hurricane damage reports, Flickr photos of the same regions, and interactive maps into a new and useful information service. A Gadget (also known as a widget) is a small autonomous program that runs visibly in a gadget container. Whereas mashups are typically executed within a Web browser, gadget containers include the PC desktop, a Web browser, or a mobile handset. A typical desktop Gadget may present a small graphical thermometer that can be configured to change dynamically according to the temperature in one's hometown, or a meter that visualizes CPU usage in an interesting way. Yahoo! Widgets and Google Gadgets directories currently host several hundreds of widgets authored by end users.

Gadgets and mashups have a profound effect on service creation:

- All aspects of service design and implementation are put into the hands of the end user/developer.

- The free and community nature creates a huge long-tail of offerings. Like other such long-tails, the great quantity does not always imply quality.

Quality and search are two key issues; the following quote reflects how Yahoo! thinks of these problems:

> "Some people have feared that making the platform too easy to develop on will dilute the quality of Widgets in our Widget Gallery. We believe that quality control is something that should be handled by the community, which brings us to our third objective: Creating the best Widget discovery and management experiences. Sorting through over 4,300 Widgets is a nice problem to have, but we still want to solve it. And the new Yahoo! Widget Gallery site and its underlying infrastructure are the solution. You'll find a bunch of new features in the site today, like improved search, ratings, reviews, tags, and our Beautiful/Useful/Fun meter..." (source: http://widgets.yahoo.net/blog/).

In Figure 4-5, analogies between the issues and concerns of mobile gadget and traditional desktop developers are presented. On the gadget side, concerns include: formulating XML, connecting to the appropriate media resources, using Web best practices, and accommodating small screens of mobile devices. The Web itself is often used as a means to host an online development environment.

In attempts to create a large user base for their own technologies, major players like Yahoo!, Google, and Microsoft have made gadget programming tools freely

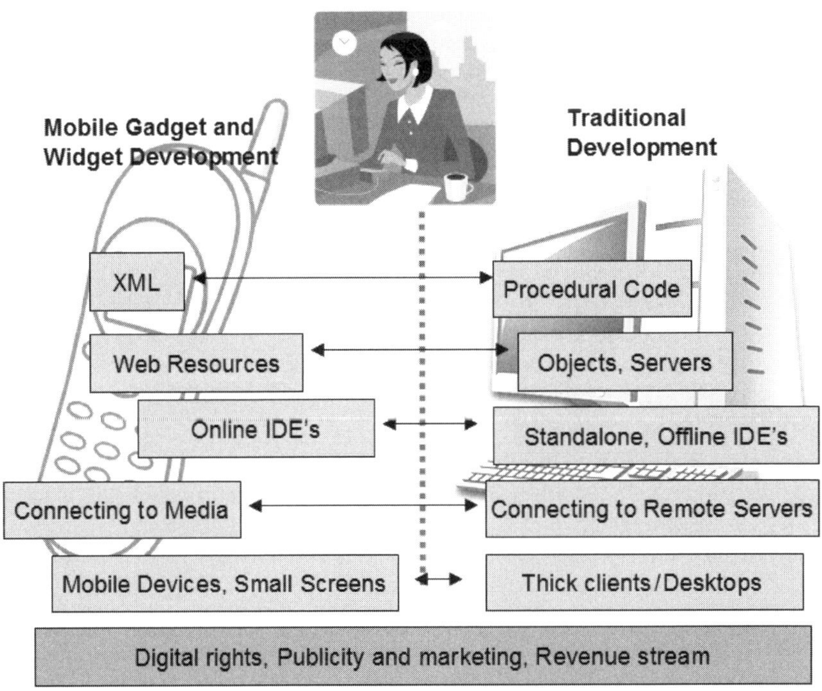

Figure 4-5: Concerns and issues: gadget development versus desktop development.

available. It is a sort of arms race in which the player with the best, most open and flexible mashup/gadget development tools will ultimately win. We note that some of these major players are also involved in service creation and delivery. They clearly recognize that instead of trying to compete with millions of developers who are willing to create and make available their own creative services, they can focus on the platform aspects and provide tools. In most cases, the tools drive the developers onto the player's Web site or otherwise create advertising opportunities. However, we don't expect this sort of tacit opportunism in future open source community-created platforms. For the traditional mobile service designer, the gadget/mashup environment is one she must learn, or risk obsolescence. Sun employee Todd Fast[7] generated buzz with a presentation at JavaOne 2008 in which he argued that size and effort are no longer the metrics with which to evaluate applications. Well-executed, small, casually developed, nontraditionally built applications, built-on "platforms" such as Facebook, for example, can become wild hits, getting millions of users in a very short time. He also points out that such applications – particularly those built in and for social networking systems – are "disposable" (meaning much more short-lived than traditional applications). Development 2.0, claims Fast, occurs at much higher levels of abstractions than traditional application development, targets the Web, and harnesses a collective intelligence. In this new world, he claims, software engineers as we know them now will shift to building and supporting the platforms, while application development jobs may be stolen from today's engineers by high school and college students.

4.3 METHODOLOGIES FOR SIMPLICITY AND USER EXPERIENCE

In this section we introduce some of the approaches to designing mobile services for simplicity and high user experience. We focus on the methods and procedures that are used by practitioners to implement simplicity and improve user experience. While this is a field with a long history, it is interesting to note how:

- Some of the classical methodologies are still relevant and now apply to mobile services.
- The nature of mobility and mobile devices is creating the need for new methodologies.

In addition to mobile service evaluation using qualitative metrics, a big part of the equation is sometimes the *fun factor* of the user experience. Companies like Apple and Nintendo are leaders in this regard; as an example, Nintendo's success with the Game Boy DS (which has sold hundreds of millions of units) shows that by focusing on fun and simplicity, it is possible to imagine your users becoming hooked on the experience that you provide [AND]. Nintendo and Apple products use controllable but consistent and predictable interfaces, three attributes that Shneiderman preaches [SHN]. Arching over consistency and predictability is

[7]http://blogs.sun.com/toddfast/

simplicity. Maeda's laws of simplicity, found in his writings on the same topic [MAED], crystallize our intuitive hunches. Some of his most relevant laws to mobile service design follow.

***Reduction*:** "Thoughtful reduction" is the process of removing any service features that are dubious in favor of simplicity. It means designers should err on the side of simplicity rather than complexity and should hide all complex features up to the point that the service still has a perceived inherent value.

***Time efficiency*:** When the user saves time, according to Maeda, it "feels like" simplicity. Time-savings in mobile services may be achieved through many tricks and tools. For example, reducing or minimizing load and wait times on screens, carefully using the working "hourglass" icon so as not to make slow tasks seem even slower, and by making necessary wait times more tolerable by providing some extra value or information for the waiting user. Intelligent caching (e.g., preloading) and background processing (e.g., doing some loading in advance or while other information is shown) are technical approaches that can help but interestingly neither of these approaches is widely used, apart from:

- Client-side scripting that reads map data and caches it so that subsequent user navigations are smooth (see Google Maps techniques).
- Layer two radio communications in which roaming handsets bind to new mobile switching centers before breaking their current association ("bind before break").

In general we expect this to change if open platforms (e.g., Linux, Android) gain momentum.

We have previously outlined user-centered design and its importance to mobile services. The key iterative steps of user needs assessment, usability testing, and user feedback are the foundations to building mobile services that give the designers the confidence to say that their service really does allow the users to accomplish real tasks. Design consistency and predictability are also important. In another design approach the IDEO design company has long touted the use of their "method card" approach: To begin thinking about a solution for a given problem, designers gather into groups and deal out method cards to each other from a deck of about 51 methods. Each card embodies an approach or spirit to design, such as empathy, emotional, behavioral sampling, day-in-the-life. The designers brainstorm solutions by forcing themselves to address the issue on the cards they are holding or combine the cards together in order to derive next steps and design decisions. For example, in playing the empathy card during design of a mobile service for elderly users, the designer is forced to think of how she could become more empathetic of the end user (e.g., by physically immobilizing or constraining parts or senses, like vision, on their own bodies!). A related approach is called "affinities diagrams" in which designers write down issues related to the design problem on small pieces of paper and proceed to iteratively group, pare, and analyze the interplay between them. The resulting organization of the concepts is a valuable starting point for subsequent design steps.

An intriguing view on mobile services design is to take into account the Gestalt laws. Applying these laws to computing is not new but Gestalt does seem ever more important in mobile computing where screen sizes are small, attention of on-the-move users is important, and the exploitation of device capabilities such as color has a pivotal impact. At a high level, Gestalt is the theory that "the whole is greater than the sum of its parts," or that the whole's behavior cannot be described as simply the composition of the parts behavior. [STU] derives an interesting analysis of the Gestalt as applied to mobile media: Understanding the user's ability to complete incomplete elements (e.g., broken shapes) allows visual features that conceal screen regions while still letting users parse the content, or, for example, correlate device's soft keys to their functions. The Gestalt laws of proximity and similarity mean that services might exploit the user's innate ability to group similar items on the screen (e.g., based on color) or to perceive objects that are close together as belonging to the same groups. These principles may come into play, for example, when designing an interactive service intended to map lots of visual information on the mobile device screen, where understanding the capabilities of the user's perception system is clearly an advantage. Exploiting innate symmetry, continuity, and figure/ground laws comprise the rest of the analysis [STU].

As mobile device screen resolution is generally smaller than that of the desktop, user experience will differ when the user views a dataset or interacts with complex information or services on one unit versus the other. From desktop-based Human Computer Interaction (HCI) research we find many – now well-entrenched – techniques, including

- Visual metaphors that embody various mental models of services and information (e.g., the computer "desktop").
- Visualization techniques that preserve and present user interface navigation context.

Providing global context while simultaneously presenting local detail is a key technique for visualizing complex information on small screens. Fisheye views and perspective walls [FUR] are the sorts of desktop approaches that will work on mobile devices as well. Even something as simple as navigating through folders or menu systems in a mobile application can leave a user confused: What are the ramifications of the current selection? Where am I with respect to the top menu option (see Chapter 7 on Design Patterns)? Zoom and scroll are fundamental as well, and the Apple iPhone, for example, does this particularly well for Web page presentation, presenting first the full view of the entire Web page and then allowing a visual zoom into whatever part of the page is of interest.

Mobile application responsiveness refers to:

- Local responsiveness – how quickly local actions are caught and responded to.
- End-to-end response time – the time between a user interaction and reception of a response from a networked resource (e.g., a search request to an engine on a network).

There are many factors that influence both of these types of responsiveness. On the local side, the applications design and algorithms are important, as well as the ability of the Operating System (OS) to handle I/O, multi-threading, and background tasks. In terms of the implementation technology, the AJAX paradigm allows for highly granular and invisible interactions between the application on the mobile device and the backend server. For example, instead of reloading a large amount of data (or the whole page) from a server each time a user interacts with an interface on the mobile device (e.g., to re-sort some tabular data), the application may have the scripts to enable resorting of the data locally (on the device), or, the application may ask the server for only for the resorted data, not the entire page structure and associated Web content (which is good because it may not have changed since the user request anyways). AJAX is the basis – at least in part – for many of today's most popular (and speediest) Web sites such as Google Gmail and YouTube.

On the server side, end-to-end response time is often hard to measure. One reason is that the insertion of metrics into a deployed mobile service may jeopardize the quality of the service itself (the "watcher" is using up resources that the watched program could be using). It is also challenging to anticipate and model the kinds of interactions the user might make with the service under the widely varying conditions of the users context (location, state of mind, etc.). One approach taken by service monitoring companies is to first manage and host a large number of network nodes across many carriers, and then as a customer needs to monitor end-to-end response time of a mobile service, to run and monitor a number of deployments of the service across these real network conditions. Careful control of the kinds of requests that the hosted service is making over the network and careful measurement of various response times can give the customer the profile she is looking for.

The W3C, steward of the standards that make the Web go around, takes mobile browsing experience very seriously. The Mobile Web Best Practices document [W3C] recognizes that, while some services are built for desktop and others for mobile browsing, it is clear that both should be available in some reasonable form on both modes. For W3C this means that an HTML/HTTP based approach should be taken. Having established this, they argue for a series of best practices grouped into the categories: Overall Behavior, Navigation and Links, Page Layout and Content, Page Definition, and User Input. The Best Practices document is a must-read for mobile Web developers; some examples of key entries include (these are paraphrased, see [W3C] for details):

- Aim to provide a balance between having a large number of navigation links on a page and the need to navigate multiple links to reach content.
- Minimize the number of images on a page and consolidate style information.
- Do not use font-related styling (since many devices have limited numbers of installed or custom fonts).
- Given the typical input limitations of a mobile device, the interface must as far as possible minimize user input.
- The balance between pagination and scrolling is partly a matter of taste and partly a matter of necessity. Devices with severe memory restrictions can only have small pages delivered to them. Equally, some devices offer a poor scrolling experience and a better page retrieval experience.

4.3.1 The Iterative Design Approach and Experience Prototyping

In service software and hardware engineering, a prototype can be an important early demonstration of the service concepts. These prototypes use approaches and shortcuts that will not transfer over to the final development process, product, or service (some of the service may be mocked-up or representative). For service developers a prototype may, for example, serve as a straw man for discussions with marketing departments or with the CTO. Should the prototype go into development, a completely different engineering approach may be taken to actually bring it to market, that is, different software platforms, different methodologies and processes, etc., but the key features demonstrated by the prototype are propagated on. A wide variety of technologies allows the rapid prototyping of services, mobile or otherwise: Dreamweaver, AJAX, Flash, HTML editors, and Mobile device emulators such as those found in the Android, NetBeans, and BlackBerry development kits. The emulators are often provided in free toolkits and are particularly useful, but developers must still be aware that features and API's supported on real hardware devices may differ from the sometimes idealized environments of the emulators.

A service prototype may be used in the context of a user interaction or user experience study in order to understand the effectiveness of the service in allowing the user to perform the tasks in question. Suri, Buchenau, and Moggridge [MOG] have written a great deal about the importance of prototyping in product and service design; they also describe "experience prototyping" (EP) in which the designers take into account the end user's contextual constraints at an untypically deep level. In more traditional prototypes, only the "concrete" sensory aspects of the service – its graphical interface, its interaction modes (buttons, mouse), its feel, its sounds, and so on – are accounted for and prototyped. In EP, social circumstance, time constraints, environmental constraints, etc. are taken into account. Thus the EP allows the designers to explore "what it might be like to engage with the product, space or system we are designing" [MOG]. EP often includes detailed role-playing by the designers in which they may go to great lengths to emulate the kind of view or interaction a typical user may have. As an example, before role-playing an elderly user to test a mobile handset designed for senior citizens, the designer may impede his own physical and attentional capabilities to better appreciate the end user. During EP service prototyping users may be asked to expose their feelings in various deep or profound ways, such as by drawing pictures or writing seemingly free form text. EP experts may exploit any of the latent information that they gather from this user output in order to improve the mobile service. [MOG] describes a use case in which test users of a prototyped online bill payment service were literally asked to "draw their money situation." EP experts then looked at the kinds of things depicted by users – an island paradise, a pile of cash beside objects that were desired – and used those drawings to help classify users into four types.

While it is possible to step through a prototype system completely on paper, paper-prototyping has its limitations. A main concern is that the user does not hold a realistic device in her hand nor does she have to push any buttons or interact in any way – rather, she engages with a design engineer who walks her through the scenarios by showing her paper mockups and describing the scenario verbally. The Wizard of

Oz testing approach is a step beyond paper prototyping. In Wizard of Oz test users interact with a physically accurate device, interface or mechanism, such as a mobile handset or computer screen. Behind the scenes while the user is interacting there is a human expert (not an algorithm!) making sure the screen shows the appropriate responses. In this way, the Wizard of Oz approach allows the testing of sophisticated scenarios that are not yet implemented but that can be driven by the "man behind the curtain." This is amusingly similar to the Mechanical Turk, an automaton of 1770 that, "appeared to be able to play a strong game of chess against a human opponent ... Publicly promoted as an automaton and given its common name based on its appearance, the Turk was in fact a mechanical illusion that allowed a human chess master hiding inside to operate the machine."[8]

4.4 CASE STUDIES: SIMPLIFYING PARADIGMS

This section describes some specific examples of current technologies and paradigms that have had an impact in mobile services for one reason or another. As technologies become increasingly complex they may become inaccessible or ineffective to both naïve and experienced users. Simplifying metaphors, new interaction types, and supporting technologies aim to fix these issues and shield users from underlying complexity; they may also shield, in some cases, mobile service developers from underlying complexity or provide a new process by which mobile services can be realized. Each of the technologies described below arrived on the scene with significant hype, and while some have realized the hype, others are now historical footnotes. We are aware that given the fast pace of development in our industry, the companies and technologies listed below will continue to evolve and may even be replaced by others yet to be invented. However, we believe that the deep innovation and know-how that are embedded in these examples hold some general validity and relevance here.

4.4.1 Simplifying Access to the Network – NTT DoCoMo i-Mode

NTT DoCoMo established a new mobile phone service in 1999 called i-Mode. For its time, although it allowed several services to be accessed via the handset, it was a breakthrough in ease of use and simplicity; at that time, other mobile handsets had very small screens capable of displaying only a couple of lines of text. The i-Mode handset began with a monochrome screen (featuring mostly Web and e-mail) but had several later more powerful incarnations that featured wildly colorful screens, graphics, and support for Java programs. Well-documented now, the DoCoMo concept was created by Mari Matsunaga, a self-confessed technology neophyte. Her main mandate was to make access to Internet and multimedia content a very simple procedure and to market the service by focusing on the use cases, not the technology. She ensured that i-Mode handsets were simple to use by employing easy to find buttons to correspond to key functions. This included a prominently placed "i-mode" button that, when pressed,

[8] *The Turk*, on Wikipedia, http://en.wikipedia.org/wiki/The_Turk

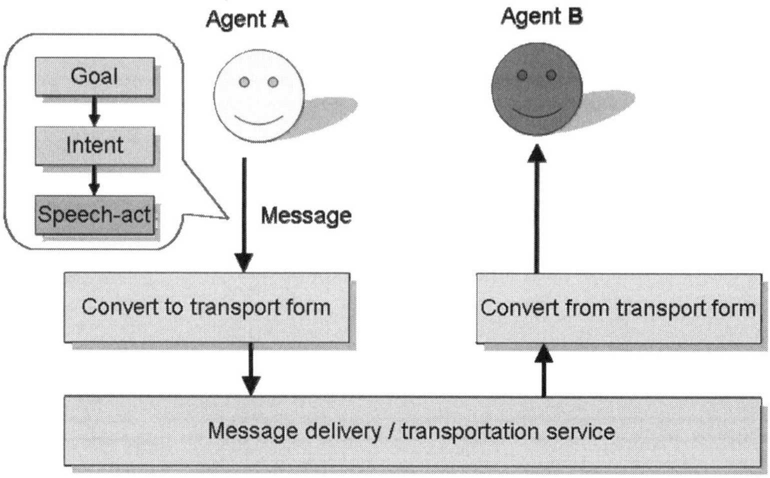

Figure 4-6: So-called "emoticons" simplify the process of messaging as they stand-in for richer sentiments. Yahoo! (left) and shareware-style (right) are depicted.

connected the user to the Internet and content. Some of Matsunaga's other design decisions were ([MOG]):

- Carve out and reserve more screen space for calendar formats (month and week views) to make "at a glance" viewing possible.
- Simplify the process of messaging by allowing users to incorporate icons into conversations. Largely to appeal to *female* users, a set of about 180 icons were defined that would support icon-based communications (e.g., a heart icon conveyed "love," a set of lips meant "kisses," etc.) This paradigm is now commonplace in mobile and Web-based chats such as Yahoo! Messenger, MSN Messenger, and so on. A free market has even arisen for cute icons that can be purchased and then reused in messaging (Fig. 4-6).
- Simplify the learning curve by incorporating well-known features from PC-based Web browsers, such as bookmarking.

Thus, NTT's i-Mode, while it did not invent each of the technologies it incorporated into its phone services, achieved something special by creating and marketing a device that both simplified and unified communications. Today, NTT i-Mode continues to bring innovative wireless service to a worldwide customer base.

4.4.2 Simplifying Access to Third-Party Content - BREW and Get It Now®

The American Service provider Qualcomm invented the Binary Runtime Environment for Wireless, a software platform that runs on a mobile device and provides support for running third-party games and applications. BREW arrived on the scene in 2001 as a low-cost mobile data delivery platform and was, at the time, comparable with other such platforms, namely J2ME and Symbian. Third-party application developers would use the Software Development Kit and a BREW emulator to develop their applications in an object-oriented language such as C++ (C language also possible). After development and test, application writers would need to upload their application to Qualcomm for digital signing and "True BREW" testing. If the application passed the testing it would be made available on the network to any BREW-enabled handset. BREW has recently evolved, however, into a broader platform that offers rich programmatic tools (e.g., uiOne) and more openness to components written by developers in languages other than C and C++.

Verizon made waves early on by branding an implementation of BREW as Get it Now®, a paradigm that made it very easy for customers to download new mobile applications. Verizon's site described it like this: "Get It Now is a technology and service that allows you to download and use applications on your Verizon Wireless Get It Now-enabled phone. For example, you can quickly and easily download ringtones, games, e-mails, directions, instant messages and more to your phone and use them right on your handset. It is a new enhancement to wireless technology." Get It Now® applications, ringtones, and music were all precertified BREW components living inside Verizon's "walled garden" of content, however, the service was very popular and shortly afterwards other operators enabled easy "few click" sequences to download Java and Symbian type components. Some criticisms of Get it Now® service included the fact that the applications were too "siloed" (i.e., available only from a few places). The BREW and Verizon stories continue to shift and play out. Regardless, Verizon's Get It Now®, supported by BREW, was an important stepping stone in the history of mobile usability and simplicity. With only a few clicks on the phones keys, even a fairly naïve end user could browse the "stack" of available content (e.g., games), easily understand the costs and options, download the content to the phone, and see the charges on the next invoice.

4.4.3 Simplifying and Enabling Hands-Free – Voice XML

In a similar time frame to BREW's introduction, Voice Web technologies also became popular as the challenge to enable easier surfing of the Web from mobile devices was answered by W3C. In 1999 they established a Voice Browser Working Group and from this group developed a suite of technologies to assist in the goal of "enabling Web developers to create applications that can be accessed via any telephone, and allowing people to interact with these applications via speech and telephone keypads." In a typical VoiceXML-enabled application the system prompts the user through a series of menu selections and user and the system take turns uttering information. A VoiceXML dialog might proceed as follows (source: W3C's VoiceXML reference pages):

Computer: *Say one of: Sports; Weather; News.*
Human: *Astrology*

Computer:	*I did not understand what you said.* *(a platform-specific default message.)*
Computer:	*Say one of: Sports; Weather; News.*
Human:	*Sports*
Computer:	*(proceeds to http://www.sports.example/start.vxml)*

VoiceXML applications greatly simplified the development of voice-enabled mobile services by cleanly separating the specification of the voice interaction scheme from the service logic that implements the service and allowing hands-free access to Web sites via spoken words and simple keystrokes.

4.4.4 Simplifying the Form - Apple

Apple's iPod designers focused on uncluttering the human machine interface and making the device do one thing – store and play music – and do it very well. The buttons on the device were all but eliminated (and hidden "behind" the click-wheel in some cases) which in turn reinforced the elegant form factor. John Maeda has pointed out that the iPod button integration has roughly followed a simplification trend with the new generations, and refers to the advantages of simplicity with the "blurring" of the iPod controls into the scroll wheel [MAED]. With the iPhone, Apple has created a killer application for the multi-touch technology. All the surfaces are clean and glossy and the messy array of buttons and keypads usually found on mobile handsets are now touchable icons on an attractive touch screen. One advantage of covering the entire front side with a touch screen is that the interface can re-assemble, adapt, and hide elements that are not relevant at a given moment, creating a user experience that is both simplified and intuitive. Noted Yale professor and author Edward Tufte has suggested that it is the (relative) high-resolution capability of iPhone's screen that is the best enabler, permitting, for example, 20 application icons to be laid out at once (or scrolled). This high-density layout allows the user to visually parse the available actions at a given moment, enabling the iPhone to focus on laying out in high density and letting the onus of computation and visual parsing fall back into the human perception system. Behind the scenes, Apple's Human Interface Guidelines gives consumers an idea of the careful thought behind the product design, particularly as it relates to simplicity; some snippets on that matter follow:

> "A mental model of a task is typically streamlined and focused on the fundamental components of the task. Although there may be myriad optional details associated with a given task, the basic components should not have to compete with the details for the user's attention."
>
> With regard to user feedback:"Provide direct, simple feedback that people can understand. For example, error messages should spell out exactly what situation caused the error."
>
> Regarding clutter: "Keep graphics simple, and use them only when they truly enhance usability. Don't overload windows and dialogs with dozens of icons or buttons. Don't

use arbitrary symbols to represent concepts . . . . The overall layout of . . . user interface elements should reflect the user's mental model of the task your application performs." (Source: Human Interface Guidelines document, http://www.apple.com)

4.4.5 Simplifying Search - Google

Google's main search page is almost completely barren. Consisting mainly of the (sometimes time-varying) logo, the input text field, and the search button, the interface is the result of some careful thinking by founder Sergey Brin. In order to focus on improving user experience Brin realized that users are put off by search sites that make it hard to find the actual search button! Brin's design reflects a conscious decision by the to de-clutter and simplify, which corresponds in principle to the company's desire to offer the best search results on the first page in the first 10 or so results. This paradigm was widely copied by other Internet search engines (both large and small). Google has made search so simple that the user now no longer even feels that she can make a mistake, which is borne out by the fact that error messages are never returned. Instead, suggestions are made in the form of questions: e.g., "Did you mean Tim Burton" might be returned on a search for "Tim Brton." Simplicity is also borne out through consistency in both the aesthetic and functional dimensions across services. Indeed, Google lists a set of "truths" that it abides by including: focusing on the user, and doing one thing and doing it well. These truths fold well into the overarching theme of simplicity in search.

4.4.6 Simplifying Mobile Development - Web 2.0 and Mobile Mashups

Web 2.0 is a suite of technologies that is all about harnessing the crowd wisdom, leveraging user-generated content, and creating a user experience better than the Web 1.0 providers like Netscape, DoubleClick, and Ofoto. Tim O'Reilly has written influentially about Web 2.0 [ORI] (see also Chapter 2). Web 1.0 applications were ultimately not delivering on the promise of rich user experiences until Javascript and AJAX technologies become much more prevalent. Nowadays those technologies make Web applications feel much more like desktop ones, especially with respect to response time. Google's Gmail interface automatically updates when new mail comes in – no manual page refresh is needed; similar streamlining is now present in all sorts of sites, pioneered by sites like Flickr, Amazon, YouTube, Zillow, Zoho, and so on. Smooth, scrolling graphics, simple page design, ample opportunity to leave and review user-comments and ratings, no page refreshes required, and other characteristics typified these sites. Google Maps, for example, achieves smooth map scrolling by silently loading map data from the server in response to how the user drags the mouse, thereby preloading adjacent regions in the map that are likely to be of interest to the user next. At the same time, Web 2.0 sites shared a simplicity aesthetic, inspired mostly by Google's simple search box. This trend was followed by almost all the mobile search portals as well as most other community and user-driven sites. While there is no rule that a Web 2.0 style mashup actually *be* simple or elegantly designed, the user experience is important. Mashups may compete with one other for traffic and poor user experience can cause a flurry of negative comments and reduce page-views.

A mobile mashup is a mashup (see earlier in this chapter and Chapter 2) that can be viewed on a mobile device. In a mobile mashup one finds several main entities: the mashup developer, a set of content providers (and many other users who may have supplied content but are not directly involved in the mashup logic), and a Web page (hosted on some Web server) that presents the mashup to the mobile viewer. Once loaded on the mobile device, the Web page is programmed in such a way as to pull content it requires from networked context providers and present it on the mobile device. By using powerful device resident technologies like JavaScript and AJAX the developer can create rich Internet applications (sometimes called RIAs) that asynchronously interwork with networked content providers and offers smooth flowing visual layouts to the user.

With Web 2.0, open programmatic access to backend data, services, and resources from Web clients became extremely important to the Web ecosystem as a whole. Note that "open" may really carry a spectrum of connotations including unlimited, rate-limited, content-limited, or time-limited. Thus, Web 2.0 sites leverage the data found on other sites in unique services of their own. Mashups *simplify* Web service creation because they allow service creators and inventors to use preexisting components and scripts, interwork with other data in the Web2.0 "cloud," and assemble a new useful dynamic service. ProgrammableWeb.com emphasizes how simple it is to get started with mashup creation by breaking it down into five steps:

- Choose what data you'd like to mashup – Is it real estate, census, historical data?
- Find a source for your data – With whom can you "partner" in order to get the data that you need?
- Consider the level of coding you need – What complexity of service are you able to support and build out? What is your development team?
- Get API keys from your sources – Enter a "business agreement" with your sources.
- Code, test, publicize – Develop, test, go live with your mashup; try to drive traffic to the site.

A final step is, of course, to advertise and socialize your mashup so that others can use it, spread the word, rate it, and provide feedback. A large percentage of today's mashups make use of mapping technology (provided by Google, Yahoo!, Microsoft, etc.) and are found in domains like travel, real estate, history, weather, and media. Following Google in mashup popularity are YouTube, Flickr, and Amazon.

Yahoo! Mobile offers a mobile development ecosystem that allows mobile applications (including mashups) to fairly easily be written, tested, and made available on a mobile portal. By making the service creation "blueprint" language both XML and Web best-practices-based (e.g., HTTP, RSS, and others) service creators can mashup a new service that pulls in feeds and media from other sites such as Twitter (e.g., user status messages) or Flickr (e.g., photo streams). The rising prevalence of mashups will continue to have a profound impact on the face of technology.

Assuming that the toolkits continue to evolve and the content continues to increase and diversify there will be no limitations as to the kinds of information services one can make with the two. Most significantly it is apparent that one does not need a computer science degree or experience in software development in order to develop a successful mashup. Open issues include:

- What are the ramifications of mashups on university curricula?
- What are the ramifications of mashups for traditional service providers?
  - The service creation lifecycle is now very short.
  - Since the Web is now the distribution channel the traditional barriers to mobile service delivery have been removed; traditional (institutional) and independent service designers can now compete on nearly equal footing; this leaves *creativity* as the main differentiator.

While the concept of code re-use and API-based access (called the "client-server" model – see Chapter 3) are not new, that a large number of often undisciplined and unschooled mashup developers are responsible for many of today's most creative services (e.g., Hurricane Katrina damage map mashup; craigslist housing-list mashup) is new and very profound. Mashup developers are proving that you need only be creative and resourceful to program a great service.

4.5 SUMMARY

We are currently in an era of amazing new mobile service possibilities brought on, in part, by ever higher user expectations. In this chapter we described a fairly large number of existing technologies and research that are directly enabling or influencing simplicity in mobile services. A popular notion of simplicity is prevalent – we see it in the literature (e.g., Maeda's writings), Blogs (Zen and simplicity sites), Website design (e.g., Google, and its emulators), technologies (e.g., Apple's iPod and iPhone, mobile mashup technologies that greatly simplify Web development), and management

Table 4-1: Summary of simplifying paradigms

Element	What It Simplified
Voice	Specification of mobile hands-free voice-driven services
i-Mode	Mobile content access over cellular
Google and Web 2.0	Mobile search and site design
Mashups	Mobile application development
Apple	Human machine interface
Get it Now!	Mobile application download
Authentication and identity management	Access and authentication with multiple mobile services
User-centered design	Resulting services are more responsive to the expectations, needs, and profiles of end users; Simple services may result.

practices (i.e., complexity in organizational structure is today seen as a drag on creativity and productivity [ASH]). Past landmark technologies have laid the groundwork for simplifying some aspects of mobile computing and services (Table 4-1).

While new networking (e.g., introduction of new radio protocols and multiple-mode mobile devices), interface (e.g., introduction of touch-screens), and profiling (e.g., multiple profiles across many services) technologies can raise the complexity level of mobile services, there is also hope that best practices, smart service design, and a move toward openness will be a unifying and simplifying influence. Having said this, the future is not entirely clear. Many feel that the so-called "mobile Web" – that is, the Web pages written explicitly for mobile device browsers and with little tie in to the "rich" Web (accessible through the desktop) – is dead. A possible paradox here: the mobile Web ends up "simplifying" (some say "dumbing down") rich Internet Web sites in order to better present them on the small screen. The results can be awkward and unappealing. Some recent studies have suggested only about 13% of overall U.S. mobile market of users actually browser the simplified Web. Improved browsers and multi-touch (e.g., Safari on the iPhone) will help on specialized devices, but access to rich, high-density sites like the New York Times will remain a challenge from a small device point of view. Perhaps the answer will lie in the creation of a mobile Web that is both appealing in style and rich enough in content to make the browsing experience worth it. On the other extreme, effective implementation of simplicity, such as how Google has made Internet search so trivial and effective, or how Web and wireless protocols can keep us informed at any given moment, can also cause us to step back and evaluate their impact. Carr proposes that Google may be making us "stupid" [CAR], while many have heard of some mobile workers becoming inextricably tied – and addicted – to maintaining mobile contact with remote people, Web sites, and media.

In the next chapter we look at what it takes to keep these sorts mobile services in "always on" mode.

QUESTIONS

1. Do you think that service creation might one day become so simple to do that even novice users will be able to define and instantiate a long-running agent-based service or a mashup that anyone can use? What technologies might comprise the underpinnings? Google Web Toolkit? Pageflakes? etc.

2. What are the most complex aspects of daily use of today's mobile handsets and how might you approach fixing them? What are the tradeoffs in cost, efficiency, size of device, etc.?

3. Not taking away seeming control is one method that designers use to empower users. Can you think of mobile services that would nonetheless serve a better user experience by taking most of the control from users, versus those that really should allow the user to remain in (seeming) control of the way the service works?

5

THE ALWAYS-ON INFRASTRUCTURE CHALLENGE

"I am sitting on a beach in Newport, but I am not entirely there. My attention is focused on a device that rests in the palm of my hand. It's a . . . smart-phone. I am using it to write this sentence, right here, right now. As a 6.2 ounce computer . . . this is one impressive micro-machine. But that is not what floats my boat. What I love about this device is its ability to reach out beyond the here and now."

—Peter Morville, *Ambient Findability*, O'Reilly, 2005.

In the preface to his book *The Victorian Internet*, which describes the history of the telegraph and the impact it had on people's lives, Tom Standage writes, "The telegraph unleashed the greatest revolution in communications since the development of the printed press. Modern Internet users are in many ways the heirs of the telegraphic tradition, which means that today we are in a unique position to understand the telegraph. And the telegraph, in turn, can give us fascinating perspective on the challenges, opportunities, and pitfalls of the Internet" [STA]. Today, smartphones and connected personal digital assistants (PDAs) are unleashing another revolution in communications that has changed the lives of many of us forever. Voice, text, and multimedia messaging (SMS, MMS), Instant Messaging (IM), and e-mail have become essential communication tools of our personal and professional lives. Similarly, the great interconnectedness users can feel through Web 2.0 social networking tools and innovative mobile services, such as those that present topical location-based information on the handset, strive to provide end users with at least the appearance of being "always on."

The Fabric of Mobile Services: Software Paradigms and Business Demands, by Shoshana Loeb, Benjamin Falchuk, and Euthimios Panagos

Like anything else, technology sometimes does not behave as intended and breakdowns do occur. Furthermore, as users, we are not quite clear about the limits of what specific technologies can offer. This is particularly true when dealing with relatively large systems that consist of many "moving parts." We can think of many examples where technical failures or our own erroneous conception of specific service properties lead to major disruptions in our lives – power failures, car not starting in the morning, online banking system down for maintenance, no cellular phone reception on our favorite mountain trail. These failures can be disruptive to those who depend on the system being always available. Imagine life for a teen without SMS for several hours. Imagine life for a prominent blogger or syndicated columnist when the service becomes unstable or goes offline. Imagine life for any practitioner who assumes that the networks and services will simply always deliver the services they need. It is not exceptional, however, for users to feel betrayed and, perhaps, place blame on the innocent when services fail. The great dependence on technology in our daily life may produce negative side effects such as emotional dependence on technology[1] or profound change in cognitive abilities [CAR].

In the previous chapter, we examined what it takes to make the experience offered by a service enjoyable and desirable for users of the service. In this chapter, we take the opposite perspective and address one of the major issues that cause user disappointment, namely, failure of a service to meet user expectations at the basic availability and performance levels.

5.1 ALWAYS-ON SERVICE PROPERTIES

Most users simply expect to be able to access online services they like at any time of the day or night (e.g., mobile banking with instant account balance transfers). These "always on" expectations are similar to what we expect of the plain old telephone service (POTS), where hearing a dial tone is always expected, even when the electricity is out. Often, however, we encounter all kinds of problems with services other than plain voice-based communications using landlines phones. Slow response times, nonresponsive operations, and cryptic error messages are common occurrences. How many times have we tried to access a specific section of a Web site and kept clicking on the browser's reload button oven and over again when the Web browser's progress bar was not moving at all?

Online services are complex and have many moving parts. First, such services may consist of several stand-alone components, and not all of these components may be available all the time. Second, users may access such services from a multitude of communications networks with different availability properties. Third, users may erroneously perceive some services as being always available and operational, even when these services do not offer such guarantees. Setting expectations and offering service disruption notifications as soon as possible are important for easing user anxiety.

[1]"Are You Addicted to Your BlackBerry?," Forbes.com, April, 2008.

We often hear about widely popular services that have experienced various degrees of slow response times or no response at all. For instance, the 2007 and 2008 outages of Research in Motion's wireless e-mail service resulted in several hours of service disruption for BlackBerry users and made front-page news. When not in the mainstream news, information about service disruptions becomes available via popular blogs or made available by Web sites dedicated to monitoring popular Web sites, such as GrabPerf.com. For instance, Figure 5-1 shows a snapshot of availability characteristics for some well-known Web sites monitored by GrabPerf.com.

Three important services properties contribute to a service being always on and available to its users: *interactive processing*, *high availability*, and *scalability*. Another important aspect of always-on services is user *perceived availability*. Even when a service is running without any problems, users may not be able to access the service due to problems associated with entities and components that are between the user's mobile device and the service itself. For example, trying to access a Web site from a data-enabled cellular phone may fail due to the failure of a Web proxy server used by the mobile operator for providing Internet access to its subscribers. In this case, the user may associate a generic error message from the mobile operator with service unavailability.

It is well known that legacy landline voice communications offer extremely high availability and reliability characteristics. In particular, landline voice services are "five nines" or 99.999% available (more on this later in this chapter). This is partly because telecommunications carriers own and operate all service components, and party because all components involved in these services have been architected with enough hardware and software redundancy to address many common failure scenarios. In the case of today's mobile networks, however, the single administrative

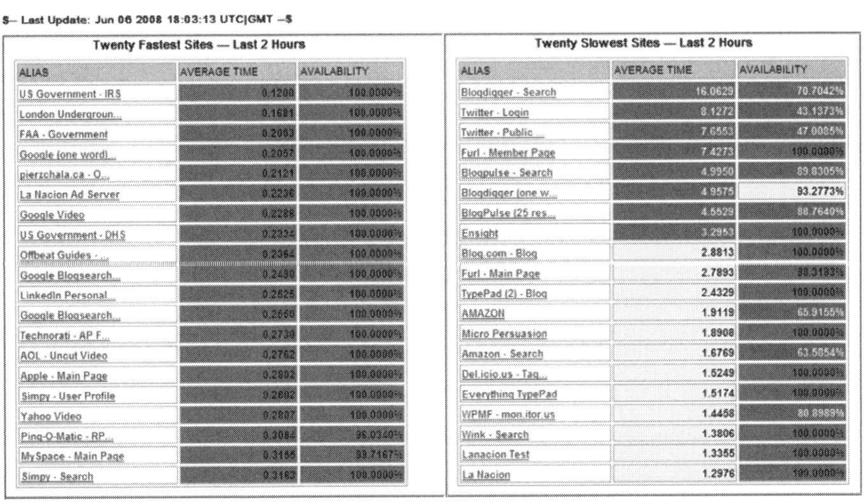
$– Last Update: Jun 06 2008 18:03:13 UTC|GMT –$

Twenty Fastest Sites — Last 2 Hours

ALIAS	AVERAGE TIME	AVAILABILITY
US Government - IRS	0.1200	100.0000%
London Undergroun...	0.1681	100.0000%
FAA - Government	0.2063	100.0000%
Google (one word)...	0.2057	100.0000%
pierzchala.ca - O...	0.2121	100.0000%
La Nacion Ad Server	0.2236	100.0000%
Google Video	0.2288	100.0000%
US Government - DHS	0.2334	100.0000%
Offbeat Guides - ...	0.2364	100.0000%
Google Blogsearch...	0.2430	100.0000%
LinkedIn Personal...	0.2625	100.0000%
Google Blogsearch...	0.2660	100.0000%
Technorati - AP F...	0.2730	100.0000%
AOL - Uncut Video	0.2762	100.0000%
Apple - Main Page	0.2802	100.0000%
Simpy - User Profile	0.2802	100.0000%
Yahoo Video	0.2807	100.0000%
Ping-O-Matic - RP...	0.3084	96.0340%
MySpace - Main Page	0.3155	99.7167%
Simpy - Search	0.3162	100.0000%

Twenty Slowest Sites — Last 2 Hours

ALIAS	AVERAGE TIME	AVAILABILITY
Blogdigger - Search	16.0629	70.7042%
Twitter - Login	8.1272	43.1373%
Twitter - Public ...	7.6553	47.0085%
Furl - Member Page	7.4273	100.0000%
Blogpulse - Search	4.9950	89.8305%
Blogdigger (one w...	4.9576	93.2773%
BlogPulse (25 res...	4.5529	88.7640%
Ensight	3.2953	100.0000%
Blog.com - Blog	2.8813	100.0000%
Furl - Main Page	2.7893	98.3193%
TypePad (2) - Blog	2.4329	100.0000%
AMAZON	1.9119	65.9155%
Micro Persuasion	1.8908	100.0000%
Amazon - Search	1.6769	63.5654%
Del.icio.us - Tag...	1.5249	100.0000%
Everything TypePad	1.5174	100.0000%
WPMF - mon.itor.us	1.4458	80.8989%
Wink - Search	1.3806	100.0000%
Lanacion Test	1.3355	100.0000%
La Nacion	1.2976	100.0000%

Figure 5-1: GrabPerf.com measurement report — June 06, 2008 18:03:13 GMT.

domain does not generally hold, and even if it did, the costs of implementing reliability would be high.

Service availability properties should not be interpreted as uptime statistics of server machines where the service is running, such as "up 342 day(s)." While such statistics are important in their own right, they do not convey the entire picture and can even be misleading in many cases. This is because services often depend on many more components than simply the server machine where they run, such as communications networks, network elements, third-party services and applications, and so on. We will talk more about this later in this chapter.

5.1.1 Interactive Processing

Traditional post-paid telephone services (landline or mobile) generate call detail records (CDRs) for all subscriber transactions (e.g., voice calls and directory assistance) and then process these CDRs in batch mode once a month in order to compute how much each subscriber owes. If a subscriber were to ask how much she owes at any time before the billing process is done with the current billing cycle, she would not be able to get an answer back. While this may be an acceptable behavior for such services, it would be a serious roadblock to many existing and future mobile services because of the element of instant satisfaction users associated with such services.

Imagine submitting a query to locate restaurants close your current location, and receiving the list of restaurants several hours or days after issuing the request. What if a picture you just captured with the embedded camera of your mobile phone and uploaded to a social networking site become available for viewing after several hours, or days? Obviously, such service behavior would be considered undesirable for an always-on service. These simple examples highlight how important it is for a service, and all of the components that make up this service, to react to user requests in real-time.

While supporting real-time service interactions for services that require limited or no interactions with other services or software applications is achievable with relative ease, the situation is quite different when complex or "dated" systems and applications, such as logistics, customer relationship management, enterprise resource planning, and fulfillment systems, participate in the service execution call flow. If any of the systems or components involved in the service execution logic cannot support real-time interactions, the service itself cannot really offer interactive processing. Imagine trying to buy a book from Amazon.com and not being able to see current book availability because the warehouse uses batch inventory management and updates the inventory database once a week.

While interactive processing is a very important property of mobile and other service categories (e.g., online services), we will not spend any more time on this subject in this book. Rather, we will assume that interactivity is present for all components that make up a particular service and focus, instead, on the availability and scalability properties of these components. In addition, we will discuss availability and scalability in the context of the server components of such services, which interact with clients on mobile devices.

5.1.2 High Availability

Hardware vendors often use the terms *Mean Time between Failures* (MTBF) and *Mean Time to Repair* (MTTR) to describe the availability of hardware components (e.g., disk drives). MTBF corresponds to the average time between failures of a particular hardware component. MTTR corresponds to the average time required to repair a failed hardware component. By using these two terms, the following equation has been used to quantified the availability of a hardware component as a percentage:

$$Availability = MTBF/(MTBF + MTTR)$$

A fundamental problem with the above equation is that it does not capture the impact of a hardware component failure on the availability of a service that depends on the failed component. For example, a disk drive failure may be repaired in a very short period by replacing the failed disk drive with a spare disk drive. However, if the failed disk drive contained data required for the proper operation of a service, such as user account information, the repair of the disk drive alone does not necessarily result in the service being operational.

To address the above problem, a more generic definition is required for being able to better quantify service availability. Once such definition is the following: *availability refers to the property of a service that describes the operational status of the service during a given measurement period.* Using this definition, the following equation computes the availability pe-centage of a service:

$$Availability = Operations\_Status\_Duration/Measurement\_Interval\_Duration$$

Typically, the duration is expressed in minutes and the measurement interval is a calendar year. The percentage computed by the above formula is commonly known as the "number of nines" property of a service. Table 5-1 provides representative examples.

To be able to attach a particular number of nines to a service, we need to be able to identify the availability characteristics of all hardware and software components that are required for the operation of this service. A service that is composed of `n` independent software and/or hardware components, with component `i` having availability a_i, has an availability of at most $a_1{}^*a_2{}^* \ldots {}^*a_n$. Consequently, the overall service

Table 5-1: Service availability expressed in terms of number of nines

Number of Nines	Description
99.0000 (2 nines)	Service not available for 5.256 minutes (87.6 hours) in a year
99.9000 (3 nines)	Service not available for 525.6 minutes (8.76 hours) in a year
99.9900 (4 nines)	Service not available for 52.56 minutes in a year
99.9990 (5 nines)	Service not available for 5.256 minutes in a year
99.9999 (6 nines)	Service not available for 31.536 seconds in a year

availability is lower than the availability of each of its components. For example, a service made up of 10 components, each with an availability of 5-nines (99.999%), has an overall availability of 4-nines (99.99%).

Labeling a service as highly available does not necessarily imply that users of the service have a high confidence in it. Here is an example that illustrates this point. Assume that a service is advertized as having 3-nines (99.9%) availability. This means that the service may not be accessible for about 520 minutes during a year. Therefore, on any given week, the service may not be available for 10 minutes. A user that interacts with the service only a few times during a week may attempt to access the service during a time when the service is not operational. If the same (unlucky) user experiences this situation a number of times during the course of several weeks, then she will probably not trust the service in the future. Thus, providers need to ensure that the service is not experiencing periodic downtime, even when the duration of the downtime is very short.

In most cases, the number of nines attached to a service does not account for *planned downtime*. Planned downtime refers to a period of time during which the service is either partially available or not available at all for the purpose of performing administrative tasks, such as patching software components and replacing faulty hardware. During planned downtime periods, users are notified in advance so that they can take any necessary steps before the service becomes unavailable. The impact of planned downtime can be minimized, or eliminated, when proper redundancy is built into the overall service infrastructure. For example, clustering of application servers allows members of a cluster to be removed from the cluster for maintenance at the expense of potential performance degradation. Ensuring that each logical component in the service infrastructure consists of multiple physical components that can be maintained independently contributes to high availability. However, there are monetary and administrative costs associated with this approach, as we will discuss later.

In many cases, availability is mistakenly confused with *uptime*. Uptime refers to a particular component being up and running. However, being up and running is not the same as being available to provide services. For instance, a software application stuck in an infinite loop may be up and running but it is not really providing any services. The same is true for two components that are deadlocked, one waiting for the other to release a required resource.

Providing high availability can be achieved by identifying all possible points of failure and managing the side effects of such failures. Typically, once the points of failure have been identified, utilization of redundant components, duplication of required information, and smart fail-over techniques for hardware/software components and data are employed for addressing single or multiple component failures. We will address the technical aspects of these techniques in subsequent sections of this chapter.

5.1.3 Scalability

Scalability refers to the performance characteristics of a software or hardware component under increased demands. Such a component is scalable when it can handle increased load in a fixed amount of time by increasing the number of resources utilized by this component in proportion to the increase in load. Under ideal conditions, increasing the available resources in a linear fashion would result in a linear increase in the load this component can handle in a fixed amount of time. This characteristic is referred to as *linear scalability*. While linear scalability is highly desirable, it is rarely achieved in practice. This is due to contention for common resources having fixed performance upper bounds, such as disk I/O, CPU, memory, network, etc.

A way to express the scalability characteristics of an application is by using the formula of the very famous Amdahl's Law [AMD], which models the relationship between the expected speed-up of the parallel execution of an algorithm relative to the serial execution of the algorithm. In particular, the following formula computes the upper bound on scalability that depends on the number of available resources and resource parallelism:

$$Scalability = 1/(\sigma + (1 - \sigma)^{*}N)$$

In the above formula, N is the number of available resources, and σ is the proportion of the service that cannot take advantage of additional resources (i.e., parallelism). We can easily rewrite the above equation in the following form, which corresponds to the *scaleup capacity* equation discussed in [GUN]:

$$Scalability = N/(1 + \sigma^{*}(N - 1))$$

For very small values of σ the scalability of an application is close to linear when the number of resources is small. However, as the number of resources increases, or the contention for these resources increases, a scalability bottleneck is reached. Figure 5-2 illustrates this for a small number of different σ values.

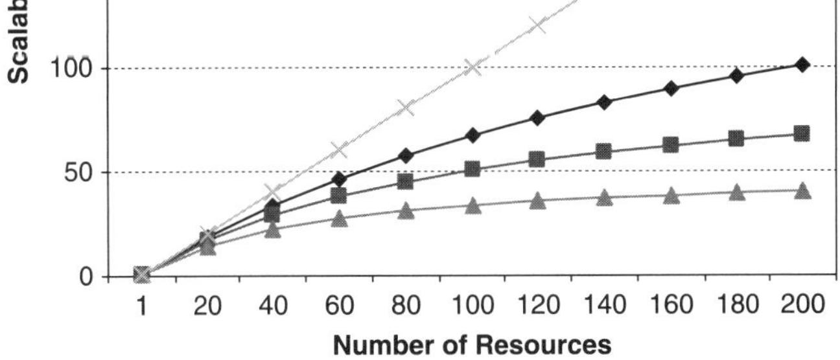

Figure 5-2: Scalability curves for different values of σ.

Several factors can limit the scalability of a service. The most important ones are the overall design of the various components of the service and the interactions among these components. A service component that has not been designed from the very beginning to handle increased load in a scalable way will reach a scalability bottleneck once the load exceeds a certain threshold. In the remainder of this section, we will elaborate on four design principles that contribute to scalability.

Synchronous/Asynchronous Communications Until recently, client-server applications (including Web-based applications and mobile applications) were designed to use a synchronous model of communication: once a request is made, the application waits until it receives a response or experiences an error (e.g., timeout or some kind of exception) before proceeding. While this approach has several advantages, including simplified logic, it may introduce unnecessary delays to service logic execution and user interfaces. This, in turn, is perceived as a scalability issue for the user. A better alternative is to use asynchronous communications so that progress can be made while waiting for a response. Multi-threaded applications, AJAX-based Web services, and event-based architectures follow this approach and allow for multiplexing of service operations.

Resource Contention The number of available resources (e.g., memory, bandwidth, disk I/O, network, data structures, and software components) and the rate at which these resources can serve requests have upper limits. Therefore, when the number of requests for a resource exceeds the rate at which requests can be served, resource contention results in increased delays. Obviously, this is a scalability bottleneck.

A typical example involves database accesses and updates to the same table row. Imagine a Web application that needs to authenticate a user against a database management system. In addition, assume that authentication includes updating the user account record with the current timestamp and IP address associated with the device (mobile or otherwise) used by the account owner. In the most common case, each application user logs into the application using a different account and only one update to a user account record takes place during a session. However, when many users use the same account, or when user account updates require updates to an auditing table, updating user accounts may become a bottleneck.

Resource Pooling Having available a number of instances of a generic object for carrying our application-specific work is the preferred way of scaling an application. Contrast this approach to creating new instances of the generic object and then throwing them away once not needed any longer.

Resource pooling is essential is scaling today's database-driven applications that require access to the database via a database connection. Without using a pool of database connections and selecting an idle instance for carrying out a database operation, the application will experience delays and, eventually, reach a scalability barrier. The reason for this is the overhead associated with the creation and destruction of database connections. Note, however, that connection pools may need some maintenance in order to keep the connections active when no application activity exists. This is because most database management systems will automatically reclaim resources not used for extended periods.

Partitioning Accessing one large database containing millions of rows in a table is not as efficient as accessing a database containing thousands of rows in the same table. There are several reasons why this is the case. For example, index structures on top of a large table become expensive to maintain due to the potential for deep hierarchies and reorganizations that may be required during inserts and deletes. In addition, accessing the references stored at the leaf level of the index requires accessing several pages containing information about interior index nodes.

Being able to partition the available resources (e.g., rows or columns in the database table) so that these resources can be accessed in parallel is of paramount importance for achieving a scalable architecture. Partitioning can be applied at several levels. In particular, one may partition resources on the same system. For example, database partitioning allows a table to be stored in multiple data files and enables indexes that are local to each partition. On the other hand, one can partition the available resources across multiple physical systems. This approach is known as the "shared nothing" approach where each system is responsible for its local resources only.

By partitioning resources across different physical systems, we can achieve great scalability and availability. When a particular system becomes overloaded, it could be partitioned further and, thus, preserve the scalability characteristics of the solution. When partitioning resources across physical systems, several issues have to be addressed with respect to accessing resources using an attribute other than the one used for partitioning. Here, one could utilize an index database for being able to locate the value of the attribute used for partitioning when some other attribute is available. Of course, the secondary attributes need to be known in advance in order to be able to lookup the primary attribute based on any of them.

5.1.4 Circuit/Packet Infrastructure Considerations

Most of today's telecommunications networks (fixed or mobile) depend on a technology designed in the late 1800s, namely *circuit switching*. In circuit switching two end-points (e.g., landline or cellular telephones) establish a dedicated and fixed bandwidth connection (referred to as channel or circuit) for the entire duration of the interaction. While this approach offers several advantages, such as simple admission control and bounded end-to-end transmission delays, it has some potential drawbacks that may affect scalability and costs associated with data-driven mobile services. In particular, circuit switching requires establishing a communications path between two end-points before any information can be exchanged. The establishment of this path requires certain amount of administrative information to be exchanged in the form of signaling between components along the communications path. As part of path establishment, state information needs to be maintained and managed at all intermediate nodes. Furthermore, when an intermediate node runs out of capacity, further path establishment requests are blocked.

In contrast to circuit switching, packet switching breaks up digital information exchanged between two end-points into a number of packets. These packets can be routed between the two end-points along different paths, without requiring end-to-end dedicated connections. Header information included in these packets is sufficient for allowing each end-point to reassemble the entire message. By being able to route packets along the best alternative route available at the time the packet is received at an intermediate node, the infrastructure is more resilient to failures and can optimize resource usage, thus offering a cost-effective and scalable solution. Furthermore, since no communication path needs to be established a priori, the connection is "always on" and the flow of data can start at any point in time.

Network operators have recognized the cost-effectiveness, scalability, and reliability properties of a packet switching infrastructure, and most of them utilize a full IP backbone these days. In addition, they have started offering IP versions of traditional circuit-switched based services, such as Voice over IP (VoIP) voice communications. Furthermore, mobile device support for multiple radios, with some of them being IP-based (e.g., Wi-Fi), is gaining a lot of momentum. By embedding IP communications into mobile devices, weak cellular coverage in some areas can be substituted with another wireless communications protocol, resulting in seamless mobile application and service operations, an important aspect of high availability.

5.2 ARCHITECTING FOR ALWAYS-ON MOBILE SERVICES

What does it mean for a mobile service to be always on? What should you do if asked to design an always-on mobile service? The initial response is to follow approaches that have worked well in the fixed computing platform case. Here, one would assume that connectivity will be there, and the TCP/IP protocol will handle any losses that may occur during periods of weak connectivity. However, this turns out to be a very naïve approach. The reality is that mobile services come in so many different flavors that it is very hard to offer one solution that fits all cases. Furthermore, the decoupling between wireless broadband operators and mobile service providers introduces additional complications that require partnerships to be established among all involved entities in to order to make the service work well. For example, Google may dictate which of its applications will be native on the Android platform. However, wireless bandwidth providers (e.g., ATT, Verizon, Sprint, etc.) may impose restrictions on these applications, such as upper limits for bandwidth used during a billing period.

In this section, we will discuss some of the technical issues and architectures associated with always-on mobile services and provide real-world examples from the areas of telematics and Mobile Virtual Network Operators (MVNOs).

5.2.1 Scalable Mobile Services and Event-Driven Architectures

The "*Worldwide PC Adoption Forecast, 2007 to 2015*" report by Cambridge, Massachusetts-based Forrester Research [YAT], forecasts that the number of personal computers will exceed 1 billion by the end of 2008 and 2 billion by 2015. While these are very high numbers, they represent only a fraction of the projected cellular phones that will be available in the same timeframe. In particular, according to market research firm Gartner, more than 1 billion cellular phones were sold in 2007, and the number of cellular phones worldwide is expected to reach 4 billion by 2010.

Given that the number of mobile handset outnumbers the number of desktop and laptop computers, mobile service scalability is of critical importance. One may think that this is not a problem since wireless network providers are already supporting large number of subscribers and keep adding new subscribers. While this is a valid point, the fact the mobile services are not necessarily owned or offered by mobile network providers raises many scalability questions. In particular, as mobile applications and services move toward an event-based architecture, being able to support a large number of events associated with mobile users is essential to offering scalable services.

In an event-driven architecture (EDA) a common middleware helps broker messages (events) sent between systems. In these special systems, event senders do not know who is subscribed to receive the events that they inject, while event receivers – once subscribed – do not necessarily know the original source of the events. These systems are bound only by the semantic of the messages they work with, and this enables a true "loose coupling" between end systems. EDA is clearly attractive for mobile systems in which a variety of information sources and data streams can be loosely coupled. Services can be – and are already – built on top of the EDA "pattern" (we will discuss this design pattern in length in Chapter 7). These would include stock services, inventory systems, and so on. Complex event-based services are also being built for enterprise and mobile spaces.

The ability to scale event systems in terms of generated events, event subscriptions, and event notifications is extremely important. Imagine a location-based service that provides the ability to remind a user when she approaches a particular geographic location (location-based reminder). To be truly scalable, such a service should be able to process location updates for its registered users and determine whether any reminders need to be sent out in real-time, and the service should be able to handle a large number of reminder notifications at any given time. Nobody would use a location-based reminder service that reminds them of things in places from which they have already taken their leave. Hence, the scalability of an event-based service depends on several factors, with the most important ones being *event filtering*, *subscription criteria evaluation*, and *event notification dispatching* [GRUB].

Event filtering is responsible for determining whether a particular event may need to be processed further in an efficient way. Being able to discard as many "uninteresting" events as possible as early as possible in the service processing logic is considered an essential property of scalable event-based services. Subscription criteria evaluation is responsible for determining whether an event matches criteria registered by event consumers (e.g., remind me 5 minutes before I reach the next post office along the current route). Here, being able to optimize event matching and perform event matching for multiple subscriptions at the same time is another critical scalability property of event-based services. Finally, event notification dispatching is responsible for dispatching matched events to consumers (e.g., sending reminders). As we already hinted above, efficient and timely event notification dispatching is crucial for service adaptation.

Event Filtering In the context of mobile services where location plays a major role, being able to efficiently determine whether a location event may need to be processed further and evaluated against subscription criteria is critical to scalable location-aware services. Obviously, the naïve approach where a location event is evaluated against all registered location-based subscription criteria (e.g., check whether current location is within 0.5 miles from a specific landmark) has poor scalability properties when either the number of such criteria or the number of location events is large, even when evaluation is carried out in parallel.

An efficient representation of location-related subscription criteria is a key factor in determining whether to both process an incoming location event further and narrow down the number of subscription criteria that may have to be evaluated against this event. Here, a hierarchical representation of location-based subscription criteria lends itself to scalable implementations. In particular, a geographic area can be split up into several regions using simple geometrical shapes (e.g., rectangles or polygons) to represent each such region. These regions correspond to the lowest level of the hierarchy, i.e., leaf level. Adjacent leaf-level regions are then combined to form higher-level regions. This process is repeated until a high-level region is formed that covers the entire geographic area.

In such a hierarchical representation of geographic regions, location-based subscription criteria are attached to the leaf-level nodes of the hierarchy. High-level nodes include an indication whether any location-based subscription criteria exist in any of the lower-level nodes located under these nodes. The root of the hierarchy includes an indication whether any location-based subscription criteria exist for the entire geographic area represented by the hierarchy. Figure 5-3 illustrates this. A set of spatial data structures (e.g., R-trees and KD-trees) exists and is well studied and well suited for this sort of region data.

When a location event arrives, the geographic area associated with this event can be determined efficiently (assuming that the underlying data structures have been used wisely). Then, the hierarchy used for representing location subscription criteria for the above geographic area is traversed to determine the criteria that need to be evaluated, if any. Obviously, if the indication at the root of the tree corresponds to no registered filters for this area, there is no need to traverse the tree since no registration exists.

The number of high-level regions to be used as well as the size in terms of geographic area covered by a leaf-level and high-level regions can be adjusted based on application characteristics. For example, for personal navigation applications, such as those offered by personal navigation device manufactures (e.g., Garmin, TomTom), the leaf nodes may correspond to specific road segments, and the root nodes may correspond to counties or states. When a particular geographic

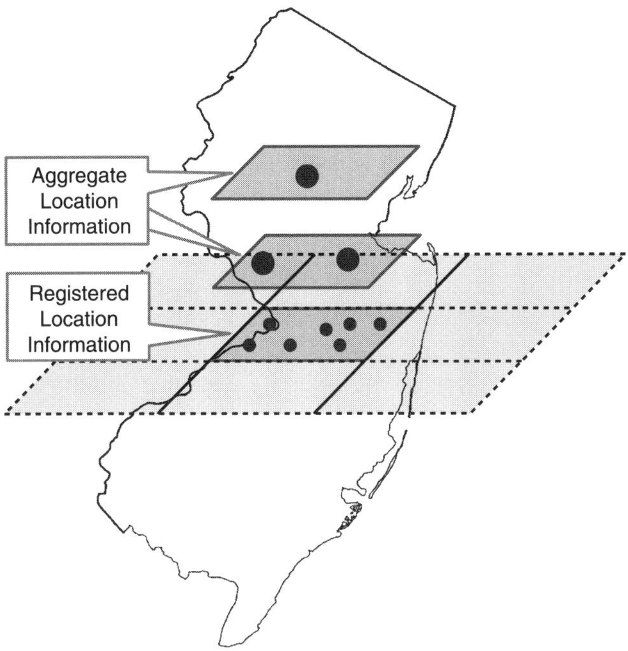

Figure 5-3: Hierarchical representation of geographic areas.

area becomes a hot spot in terms of either the number of location events associated with this area or the number of location-based subscription criteria, the area may be split into two or more sub-areas and each new area may be assigned to a different cluster of servers in order to maintain application scalability. The computational complexity of restructuring spatial trees is also well studied.

Equally important to efficient location representation is temporal representation. For example, imagine a personal navigation application that is integrated with a user's calendar and offers turn-by-turn direction calculations for the location associated with a calendar appointment based on traffic conditions along the route. Clearly, an accident along the selected route would affect travel time and, hence, appointment timeliness. However, if the accident occurs many hours or days before the appointment's start time, there may be no impact on travel time. In this scenario, determining whether a particular event (e.g., traffic accident) needs to be processed further based on whether it may affect time-related subscription criteria is a key factor in overall scalability.

Similarly to the hierarchical representation of geographic areas, a hierarchical representation can also be used for time-related information. Figure 5-4 illustrates at a high level how one could efficiently represent time information in a hierarchical way that offers a scalable solution to determining whether there is an event or interest for a particular point in time. In this representation, a four-level hierarchical structure for encoding events associated with particular times is used. Here, the first level corresponds to a calendar year. The second level corresponds to a month within a calendar

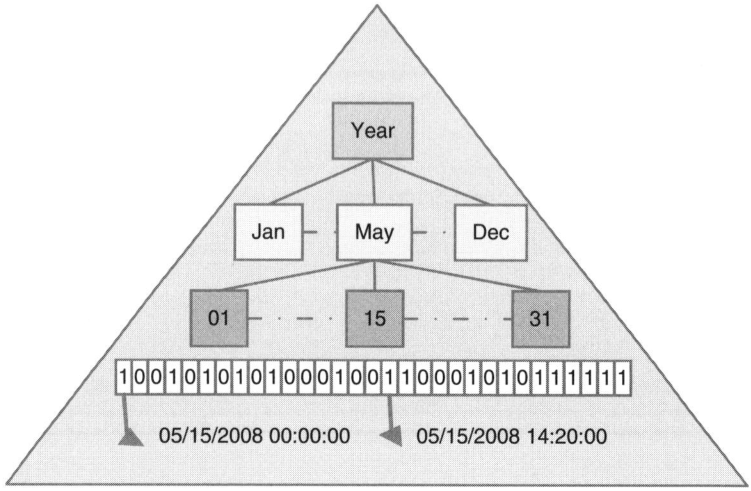

Figure 5-4: Hierarchical representation of time information.

year. The third level corresponds to the day of the month. Finally, the fourth level encodes time of day information.

In order to efficiently encode time of day information, we divide a 24-hour day into a number of **n** minute intervals, where **n** can be as low as 1 and as high as 1440 (number of minutes in a 24-hour day). Typically, a value of 10 or 15 is adequate. By utilizing a bitmap to present a time interval and setting the bit to 1 when at least one registered entry exists for the interval in question, we can efficiently determine whether an entry exists for a particular time of day, month, and year, by just performing bitmap operations, which are extremely efficient. Assuming a value of **n** equal to 15, we need around 4 kbytes of storage for implementing the above-mentioned bitmap for an entire year. The above temporal and spatial data structures are purely illustrative but show how with the correct approach, event filtering performance may be improved.

Subscription Filter Evaluation Subscription filter evaluation is CPU intensive and event dispatching is I/O intensive. Therefore, overlapping them would increase parallelism and contribute to the scalability of an event-based service. This observation suggests that separate processing threads could be used for filter evaluation and event dispatching. However, in order for this approach to scale, the overhead of the extra processing threads and synchronization that may be required has to be less than the overhead of having the same set of processing threads responsible for both filter evaluation and event dispatching.

There are two main dimensions for efficient subscription filter evaluation: parallel processing of the events generated by suppliers, and parallel processing of the filters registered by consumers. We explore each dimension separately, assuming sequential execution of the other dimension. Under parallel processing of events, a single thread evaluates all consumer filters. An obvious benefit of this approach is that no synchronization is required at the event level. However, two or more threads for different events may evaluate the same filter concurrently. Therefore, when each filter maintains internal state for each evaluation, synchronization may be required for protecting the consistency of such state. This synchronization overhead may be substantial when a large number of filters are used and, furthermore, events arrive at a high rate.

Depending on the complexity and selectivity of filters, the evaluation of all relevant filters for a given event may be completed faster than the evaluation of all relevant filters for a second event by a different thread. Therefore, the order in which events are delivered to consumers may differ from the order in which these events were supplied. While this is not an issue for consumers that are not sensitive to event ordering, it introduces additional overhead when events must be dispatched in the order in which they arrive (most monitoring applications require events originating from the same supplier to be dispatched in the order in which they were sent to the channel).

Under parallel filter processing, disjoint groups of filters are evaluated in parallel for the same event by several threads. Here, different criteria can be used for grouping "related" filters together. Since each filter is accessed by one thread, synchronization requirements at the filter level are not required. Furthermore, there is no mismatch between event arrival and filter evaluation order because each event is processed after filter evaluation has taken place for all events that arrived before it. Finally, the number of threads and filter groups can reflect the complexity and number of registered filters, leading to better scalability.

However, because an event may be accessed by many threads at the same time, updates to event-specific state would have to be serialized, or each thread would have to maintain its own copy of event state. Event state is needed when filters specify constraints that need to extract specific value parts of event attributes. Furthermore, the allocation of filters to filter groups may lead to substantial differences in the processing overhead for each such group or unnecessary evaluation of filters for the same event.

Event Dispatching Events are dispatched to consumers after the successful evaluation of the filters registered by them. Since the event service is responsible for sending matched events to consumers when push communication is used, the performance of the event service depends on the dispatching strategy employed. Assuming a large number of consumers, a multi-threaded event dispatching architecture is the preferred choice. The two main threading alternatives here are *thread per consumer* and a *pool of threads*.

Under the first choice, a thread is created for each consumer to handle the forwarding of events. Although this approach is simple to implement, it can consume a large number of local and network resources when the number of consumers is high. In addition, it is very inefficient for consumers that connect to the service for short periods, as it may be the case for mobile devices. Finally, this approach many not be practical for some notification channels, such as SMS. This is because the number of applications connected to an SMS delivery platform at the same time is typically limited to a very small number.

In the thread pool approach, a fixed number of threads are used for dispatching events to all connected consumers. This approach avoids the shortcoming of the thread per consumer approach since these threads are *not* associated with a single consumer. However, it may require synchronization between threads when they share the same queue(s) that contain the events that need to be dispatched.

5.2.2 High Availability for Mobile Services

Due to the global nature of mobile services, being able to access a service at any point in time and from any location requires an infrastructure that can mask failures without any noticeable service degradation. Several well-publicized instances of service blackouts are available. For example, RIM's BlackBerry email service has had several occasions where service was not available for quite some period. In several of these cases, the problem was with the infrastructure managed by RIM and not the cellular provider. What this implies is that it is very important to examine all components participating in a mobile service and ensure that these components can handle as many failure scenarios as possible.

Let us look at a specific mobile service in the area of telematics. Telematics services require a highly available infrastructure for offering safety-related services, such as accident notification with airbag deployment. Telematics services are being offered via a combination of hardware, software, and telecommunications infrastructure. On the hardware side, a telematics device is installed in a vehicle. This device, referred to as On-Board Unit (OBU) or Telematics Control Unit (TCU), serves as the gateway for all interactions between Telematics services and a vehicle. Typically, the OBU is connected to the vehicle network (CAN bus), contains a GPS receiver, and one of more radios to be used for network communications.

On the software side, software running on the OBU device is responsible for implementing specific functionality associated with the telematics services enabled for this device, such as remote diagnostics, vehicle health, geo-fencing, and remote door lock/unlock. In addition, OBU software is responsible for message exchanges with the back-end telematics infrastructure by using one of the available radios and the most appropriate communications channel (e.g., SMS, data). OBU software runs on top of an operating system (OS) installed on the device, and it utilizes resources available by the device, such as persistent memory (EEPROM). Figure 5-5 shows the high-level architecture of a typical telematics service offering.

In Figure 5-5, an OBU connects via the cellular network to telematics services. Device gateways are responsible for interacting with devices via IP connections, SMS messages, or other communications protocols, standard or proprietary. Device gateways interact with application servers using standard Internet protocols, such as TCP/IP and XML. Application servers are responsible for implementing the business logic associated with the telematics services being offered to consumers. Application servers interact with external content and information providers and, in addition, they are accessing persistent information stored in relational databases.

Given the relatively large number of components involved in offering telematics services and, in addition, the dependency on network operators for providing connectivity between the OBU and the back-end telematics infrastructure, offering high-availability is not an easy task. Various hardware and software errors need to be

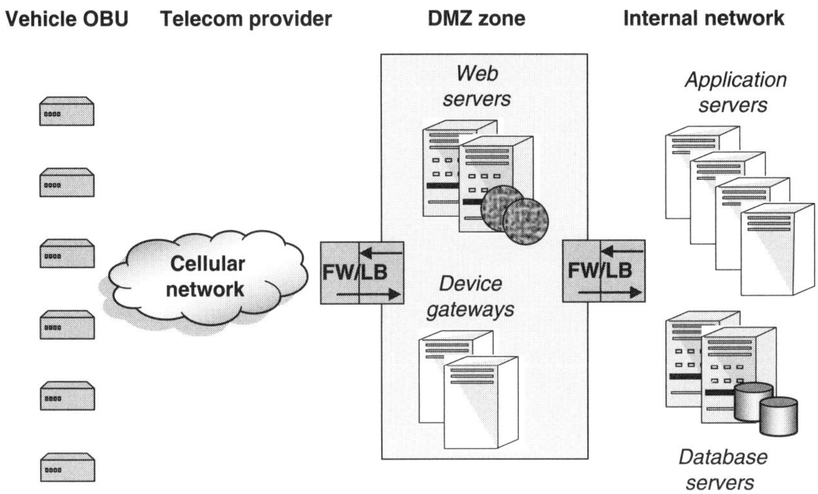

Figure 5-5: High-level telematics service architecture.

addressed at various points in the end-to-end delivery of services, together with potential communications problems. Since well-known approaches exist for handling hardware failures, such as redundant components (e.g., dual power supplies) and clusters of servers with traffic load balancing, we will not discuss such failures further. Rather, we will concentrate on key software design principles that contribute to high availability service properties.

Reliable Message Exchanges Message exchanges between an OBU and the back-end telematics service must be reliable. For instance, accident notification with airbag deployment messages must be reliably received by telematics services as long as the OBU has access to at least one communications channel. Depending on the message delivery guarantee properties of the communications channel used for message exchanges, different approaches may need to be incorporated into the service for ensuring reliable message delivery. In addition, explicit message acknowledgments may be required, especially when asynchronous messages are used (e.g., SMS messages).

Reliable message exchanges can be achieved by following a few simple design principles. First, sent messages should be remembered until they are not needed any longer (e.g., after they are acknowledged). Second, messages should be uniquely identified in order to be able to handle duplicate messages without any undesirable side effects. Third, timeouts should be used for all asynchronous message exchanges. When either a reply or acknowledgment to a sent message is not received within a specific time period (should be configurable and depend on service use case), then the message should be re-sent. A configurable number of re-tries should be supported in order to handle the semantics of the operation associated with the message.

Multi-channel Communications When several communications channels are available, such as voice, SMS, cellular data, Bluetooth, and Wi-Fi, then channel selection for message exchanges must be driven by policies associated with the services offered. The available policies must be designed by considering the following criteria:

- *Communication Channel Priority*: In the presence of multiple communications channels, priorities should be assigned to these channels for each class of message exchanges. Channel priorities can be driven by channel reliability and bandwidth properties, as well as any costs that may be associated with using this channel.
- *Communication Channel Availability*: If a specific communications channel is not available or experiences weak connectivity (e.g., no cellular data connection or weak signal), then an alternative communications channel should be selected. The channel selection should be based on the above-mentioned channel priorities.

Exception Handling The execution flow of a service may require interactions with multiple components, such as auditing, persistent storage, and billing record creation. One should not assume that all these components would always be available. Therefore, the service should be designed to cope with unavailability of some of these resources, as long as these resources are not critical in providing the service functionality. What this means to service design is that interactions with components that are not crucial to service functionality should be optional under certain conditions, such as when these components do not respond within a particular time period or when the overall service load exceeds a threshold value.

While we focused on the telematics space during our discussion about highly available mobile services, the software design principles we presented are applicable to any mobile service that requires message exchanges with a network component.

5.2.3 Always-on System Architecture Considerations

In the previous two sections, we focused on scalability and availability properties of mobile services by analyzing the scalability of event-matching components, a very important class of components for mobile services, and discussing software design principles for highly available mobile services that depend on message exchanges between a mobile device and a remote application. In this section, we will take a step back and focus on the high-level system architecture of scalable and available mobile services. To make our discussion concrete, we will focus on the architecture used by many telecommunications service providers for offering advanced intelligent network services for mobile subscribers, such as those being offered by today's mobile network operators (MNOs) and mobile virtual network operators (MVNOs).

Advanced mobile services depend on persistent data and service logic. The service logic executes in software, and it is driven by a combination of synchronous and asynchronous message exchanges with the core telecommunications network over dedicated links – Signaling System 7 links (SS7). Information about which service features are enabled for different classes of subscribers, including subscriber balance information for pre-paid offerings, is stored in persistent storage.

Because these services are tied directly to the revenue stream, due to either fees associated with prepaid subscriber transactions, such as voice calls, SMS messages, and data downloads, or reduced subscriber churn rates, offering a highly available service is critical to business operations. Consequently, several instances of the service execution logic and associated persistent data are used. Furthermore, to address natural disasters and other large-scale problems that may affect service operations, service instances are geographically distributed to at least two physical locations (sites). Within each site, several service instances are active for addressing hardware/software faults and upgrades transparently. Figure 5-6 illustrates the typical service architecture.

Because service sites are connected via a Wide-Area Network (WAN), which is characterized by limited and expensive bandwidth, the overall system architecture must optimize WAN traffic to avoid both costly WAN connectivity and increases in latency. In addition, to ensure that overall service performance is not affected when one of the service sites becomes unavailable and all traffic is handled by the remaining sites, each service site is configured to operate at less than 50% capacity under normal conditions.

While any service instance can execute service logic on behalf of any subscriber, subscribers are typically assigned to one of the available sites. This is typically achieved by specifying the Area of Service (AOS) a subscriber is assigned to in the subscriber's Home Location Registry (HLR) record. The AOS corresponds to one of the sites. There are several reasons for doing this. Since the vast majority of subscribers exhibit similar calling patterns, assigning subscribers to specific sites achieves

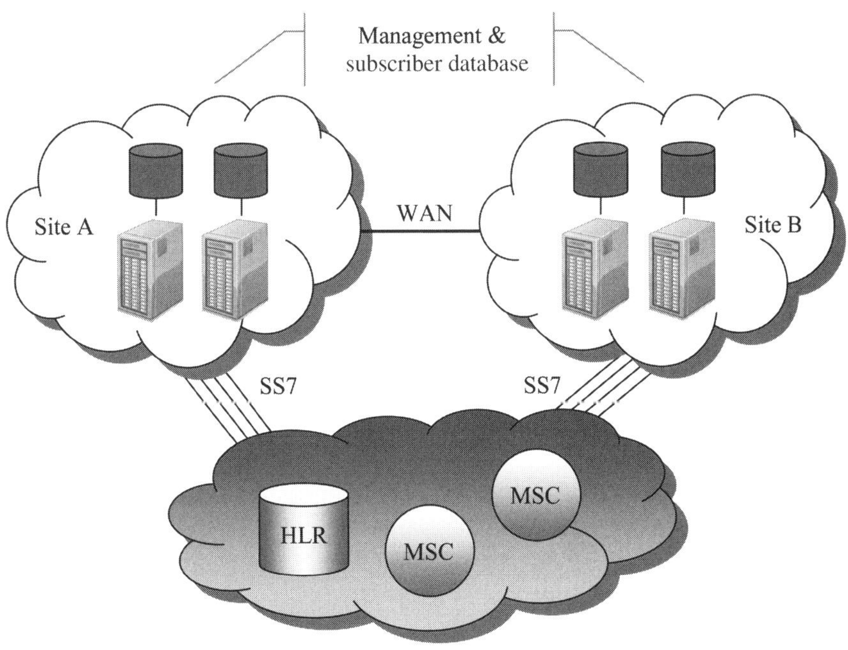

Figure 5-6: Typical advanced intelligent service system architecture.

load balancing between sites. Furthermore, session state information related to specific subscriber transactions is kept at the same site and, hence, there is no need to replicate such state to other sites.

In order to address failures that affect an entire site, subscriber data is replicated between sites and, in some cases, among service instances within the same a site. This is done in order to be able to continue offering services to subscribers whose primary site becomes unavailable. At each site, subscriber data can be stored in a shared database that is accessible to all service logic execution instances, or, alternatively, each server machine may have its own copy of the subscriber database (shared-nothing architecture). Each approach has its own merits and drawbacks. In the shared database approach, there is only one copy of subscriber data per site and, hence, no inter-site replication is required. However, a software or hardware error with one of the service instances or server machines may make the shared database inaccessible, affecting overall service availability.

In the case of prepaid services, the subscriber database experiences high update rates, in the order of tens of thousands of updates per second in some cases, because subscriber balances need to be credited and debited in real-time. Therefore, the database must be able to support such high update rate. Existing commercially available databases could be made to scale and handle this above large number of update transactions at the expense of specialized hardware (e.g., solid-state disks) or a large number of server machines. Both of these approaches may prove to be quite expensive, especially when only a small percentage of the available database functionality is being utilized.

Typically, main-memory databases are being used for achieving the performance requirements of such services with minimal costs. In addition, since prepaid solutions deal with debits and credits to subscriber balances and these operations can be executed in any order without affecting the result, simplified main-memory databases are being used (commercial, open-source, or proprietary).

For scalability, each service deployment (service instances, service machines, sites, and subscriber database) is sized to handle a specific number of subscribers. When the number of subscribers increases beyond the maximum number that can be handled by existing service deployments, an additional service deployment is required. This service deployment will have its own subscriber database, which will contain subscribers that do not appear in any other subscriber database used by existing deployments. Since there is no need for synchronization or any form of coordination among service deployments, linear service scalability can be achieved.

The availability and scalability solutions we discussed so far in the context of advanced intelligent network services provide valuable lessons for many large-scale mobile services. The following list summarizes the most important properties of highly available and scalable architectures:

- Multiple service execution instances distributed in different geographic locations;
- Intelligent load balancing of service requests across available service instances to avoid service state replication;

- Groups of service execution instances responsible for managing the same service data (user profiles, multi-media content, etc.);
- Service data partitioning across groups of service execution instances following a shared-nothing approach.

5.2.4 Virtualization

Virtualization is a software technology that transforms hardware resources of computers and mobile devices (e.g., CPU, RAM, storage, network controllers) into fully functional virtual servers that can run their own operating system and applications in the same way as the original computer or mobile device. By doing so, virtualization enables efficient and secure sharing of physical resources across multiple operational environments, making the original computer function as several independent computers.

Virtualization technology was first developed in the 1960s for mainframe computers out of the necessity to fully utilize the expensive resources (in terms of dollar amounts) of such computes and be able to run multiple applications and processes at the same time. In the 1980s and 1990s, virtualization was overshadowed by client-server and distributed computing based on inexpensive desktops and servers. Today, virtualization is experiencing a renaissance due to the high infrastructure and management costs and low utilization of the plethora of available desktops and servers.

Since virtualization allows the creation of virtual machines where operating systems and application can run, several underutilized servers could be represented as virtual machines running on a single server. By doing so, several of the existing servers could be consolidated into a small number of machines resulting in considerable saving in hardware costs, data center costs, management and administration. Furthermore, the ease at which virtual machines can be created and installed on different servers contributes to improved application availability, one of the main always-on components.

Existing commercial, open-source, and research virtualization solutions provide an infrastructure where virtual servers are constantly monitored and are restarted on different physical servers when a failure is detected at the physical server level or, in some cases, virtual server level. Some of these offerings allow migration of virtual servers with minimal or no service interruptions [CUL]. Because of these features, virtualization offers a software-based application agnostic high availability solution without relying on specialized hardware resources or special-purpose software. Figure 5-7 illustrates a virtual server migration scenario when one of the three available physical servers fails.

While most of us consider virtualization to be an important technology for back-end and infrastructure components, virtualization is poised to play an important role in the future of mobile devices and applications being offered for such devices. The "Open and Secure Terminal Initiative (OSTI) Architecture Specification" by Intel and NTT DoCoMo provides details on how virtualization can be applied in the mobile handset space [INT06].

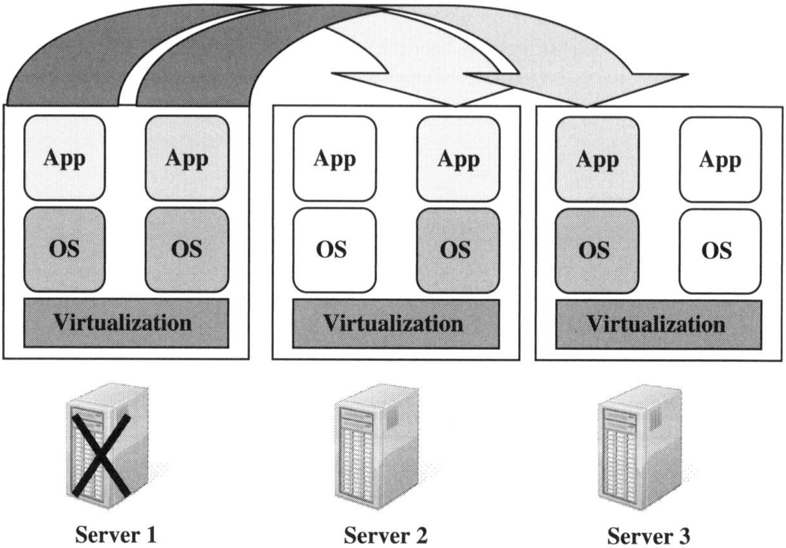

Figure 5-7: Virtual server migration example.

The OSTI architecture is geared towards enabling application, content, and information providers to offer value-added services on a cellular phone without compromising the quality and reliability of Operator services. In particular, the OSTI platform allows enterprise applications to co-exist with traditional Operator services (e.g., voice, SMS, MMS, content and information access) on a mobile device. Furthermore, enterprise applications have their own operating system and GUI.

Figure 5-8 shows the OSTI platform architecture. In this architecture, the Domain Abstraction Layer (DAL) provides a standard interface between the Enterprise OS and the rest of the platform.

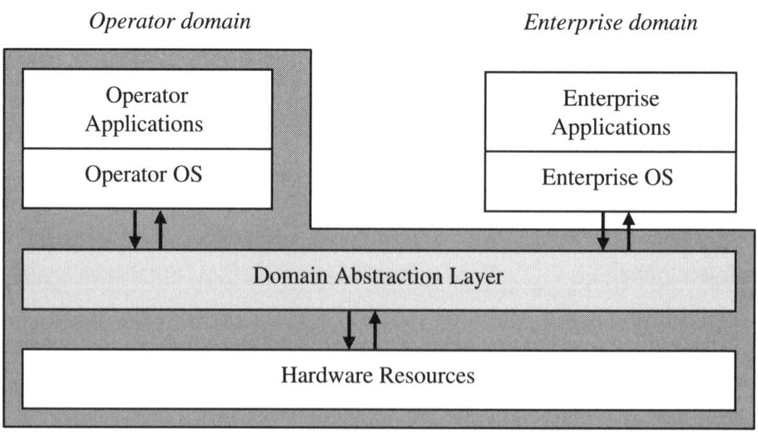

Figure 5-8: OSTI Platform Architecture.

OSTI depends on two technologies that allow two operating systems to share the same hardware resources available on a mobile handset: OS switching and virtualization. OS switching allows users to switch between the two modes of operation (Operator and Enterprise) in a very efficient way by exploiting the "sleep" mode supported by modern operating systems for conserving power. Virtualization allows the co-existence of multiple virtual containers (operating system and applications) on the same device.

While the OSTI architecture focuses on just two domains, one can envision multiple such domains being available on the same mobile device. Depending on the available hardware architecture (e.g., dual-core CPUs, long-lasting power sources), the various domains available on the device may be active at the same time. Furthermore, users may be able to create temporary domains for running specific applications, such as non-trusted applications downloaded from the Internet.

5.3 BUSINESS IMPACT OF ALWAYS-ON

In this section, we will discuss some of the business impacts of offering always-on services. In particular, we will discuss how business decisions may sometimes negatively affect user expectations, how proper partnerships and business agreements may affect both service functionality and pricing, and how not being prepared to handle increased service usage in an efficient way may make a service obsolete.

5.3.1 User Expectations

No matter what the service characteristics are, user expectations are always hard to match. Although users are accustomed to imperfect cellular connections (how many times have we found ourselves saying "hello" to make sure that the connection is still there and the other party can hear us?), when accessing mobile services that offer transactional guarantees, such as mobile banking, they will not tolerate slow or non-responsive service. The same is true for turn-by-turn directions or any other services that have timeliness constraints. What if the service becomes unavailable for an extended period while driving and it is not able to retrieve updated instructions?

Another subtle issue with wireless broadband is usage limits attached to data traffic. In many cases, mobile operators impose an average limit of between 3 and 7 gigabytes per user per month. Users who exceed the limit may experience service interruptions and even service termination. With mobile handsets being able to receive streaming audio and video, as well as produce streaming audio and video, the imposed limit can be easily exceeded. In such cases, users will most certainly be taken by surprise.

A recent example is that of Verizon Wireless. In 2007, Verizon Wireless changed the terms of their unlimited wireless broadband data service. Under the new terms, users exceeding 5 gigabytes of data transfers per month would be *penalized* by having data rate restrictions imposed on their connections. These data rate restrictions would downgrade the connection speed down to less than 56 kilobits per second.

Clearly, this is not what users would have expected under an unlimited wireless broadband service.

The above example highlights a very serious issue with wireless broadband data connectivity and user expectations. Since voice capacity is sacrificed for offering data services, wireless operators will be unwilling to allow unlimited data access as long as the majority of their revenue is generated by voice, SMS, ringtone and wallpaper downloads, and walled garden services. Therefore, users expecting to access streaming audio and video, engage in peer-to-peer content sharing applications, VoIP sessions, etc. will be surprised when some of these services will be denied or perform badly after usage exceeds some threshold.

5.3.2 Partnerships and Business Agreements

Mobile services depend on four main components: content providers, service providers, mobile device manufactures, and wireless network operators. Content providers make content such as audio and video available for download to mobile devices. Service providers offer services such as traffic and weather information to users. Mobile device manufacturers offer advanced functionality and computing infrastructure on mobile devices. Finally, wireless network operators offer wireless broadband connectivity so that users can access content and services using their cellular phones or PDAs. What is interesting about the above component is that the parties involved may be both customers of and competitors with each other. Figure 5-9 illustrates these complex relationships.

According to a recent report by IDC [IDC08], mobile device manufacturers are facing an uncertain future in the next several years as the mobile telephone growth

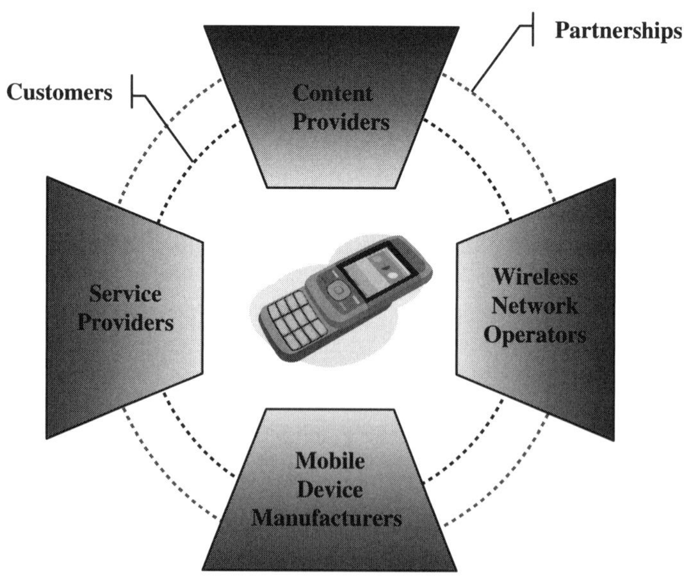

Figure 5-9: Relationships among mobile service players.

is expected to be at 5.8 percent. This is a steep decline compared to the compound annual growth rate of 20 percent from 2001 through 2007 reported in [IBM07]. Less growth and enduring price pressures have an impact on the bottom lines for mobile device manufacturers. Since mobile device technology has become a commodity, mobile device manufacturers have started to pay attention to user interfaces and offerings of mobile services in order to differentiate themselves, as exemplified by the introduction of Apple's iPhone in 2007.

Similar to mobile device manufacturers, mobile network operators are facing their own battles for growing their market share in a profitable way. While the current battle focuses on wireless broadband speed and coverage, offering unique and engaging services on top of high-speed (3G and 4G) wireless networks will be the ticket to staying competitive. A 2008 survey by IBM Institute for Business Value shows that consumers prefer to receive mobile applications and services from a single provider instead of having to access these applications and services from independent providers over the wireless broadband connection offered by their wireless service provider [IBM08].

Given the synergistic and, at the same time, competitive nature of the interactions among the players involved in the delivery of a mobile service, mobile service providers face some tough decisions in making their services available to users. The following list provides details about the most important of these decisions:

- *Dedicated Mobile Device*: Should a dedicated mobile device be used for delivering the service or could existing consumer electronics devices be used for this purpose? The choice has implications on the available functionality, quality of the service, and cost of the service to both consumers and service providers. Let us look at two specific examples from the telematics space, GM's OnStar[2] and Ford's SYNC.[3] OnStar service follows the dedicated device approach by integrating a CDMA mobile phone in vehicles for providing communications functionality with network resources and call centers. SYNC, on the other hand, depends on a consumer mobile phone for the same purpose. In the case of OnStar, a separate service plan is required with a mobile network operator for the in-vehicle cellular device (Verizon Wireless in the United States). Service charges associated with this plan are incorporated into the monthly fees paid by service subscribers. In the case of SYNC, no separate service plan is required, resulting in lower prices for the same services.
- *Integrated Internet Access*: Should Internet access be offered directly from the mobile device or should all information and content accesses be carried out using a dedicated service portal? The choice has implications on the available selection of content and information accessible by the mobile device, as well as the quality of the available content and information. For example, allowing mobile devices to access the Internet directly provides a lot of flexibility. However, not all

[2]See http://www.onstar.com for details.
[3]See http://www.syncmyride.com for details.

available Internet content and information is optimized for viewing on mobile devices.

- *Service Fees*: Depending on the elections made for the above two categories, different services fees may need to be applied. In particular, when the consumer's mobile device is used for accessing the service, then the service itself does not incur any communications costs. Therefore, the service could be offered at a lower price. Furthermore, depending on the predominant communications pattern between a mobile device and the mobile service, a service plan optimized for the most common case may be selected. For example, a telematics service that is driven by mostly data exchanges with an OBU may use a cellular data plan only. If a voice channel is required for short and infrequent voice communications, a VoIP solution might be used on top of the data plan.

5.3.3 Over/Under Engineering

As many start-up and well-established companies have experienced over time, there are business costs associated with both over-engineering and under-engineering a solution. At one extreme, one takes the very optimistic view that the release of a new service would be so successful that millions of users are going to access the service over a short period. Therefore, the service must be extremely available and the infrastructure should be able to support the traffic. This implies a large number of hardware boxes and software licenses for middleware and database products, as well as expensive data center hosting. As a result, the company may spend a lot of precious money in setting up a highly scalable and available infrastructure that may not be fully utilized until several months or even years after the launch of the service. This approach works well when the new service sees rapid and large growth.

On the other hand, one may take a minimalistic approach and deploy just enough resources to handle a moderate amount of traffic and plan to add additional resources as the service becomes more popular in a gradual manner. In several cases, however, a service becomes so popular in a very short period of time that results in slow service response or no response at all for prolonged periods. In this case, if the scalability and availability of the service are not addressed quickly, its users will abandon the service and, hence, the company's reputation will be tarnished. This approach works well when the new services sees slow to moderate but steady growth.

The choices made in both the overall architecture and technologies used for implementing a service have an enormous impact on the scalability and availability properties of the service, no matter how much additional hardware resources may be thrown at the problem. Take, for example, a service that uses a middleware platform that simplifies service implementation at the expense of restricting how persistent resources can be accessed. When persistent resources, such as databases, become a bottleneck, additional resources may be used for alleviating the bottleneck. However, if these additional resources cannot be used by the middleware platform, then these resources would have no impact on the scalability and availability of the service.

5.4 DATA MANAGEMENT

There are two main classes of data associated with users of mobile services: mostly static, and dynamic. Examples of mostly static data include user preferences, service policies, billing plans, and application data such as contacts lists. Examples of dynamic data include calendar application data, social networks, location information, current user activity, SMS messages, and content captured by the mobile device, such as images and video clips. Managing such content presents several challenges in terms of which data model to use, how to make the content highly available, and how to provide efficient access to it.

While a relational database is an acceptable solution for storing mostly static data, dynamic information storage is quite challenging. This is due to both the potential high volume of such data and the queries one may want to ask against such data. Partitioning and replication are well-known solutions to scaling mostly static data. However, scaling dynamic data may be quite challenging in the presence of large number of updates, such as current location information. In the remainder of this section, we will touch upon the problem of how to model location information in order to support location-based queries in the presence of missing data and trajectories. In addition, we will discuss how partitioning and replication can be applied to dynamic data.

5.4.1 Modeling Location Information

A major capability of a location-aware service is modeling of transient location information as this information is either pushed by or pulled from a mobile device (cellular phone, telematics control unit in vehicle, etc.). Typically, such information is represented as a pair of location and time values. The location value is extracted from either a GPS receiver associated with the mobile device or telecom infrastructure utilized by the device (triangulation of cell towers, bases stations, etc.). The time value is extracted from either the GPS receiver, local clock, or a network clock.

The straightforward approach to organizing location information is to store a (location, time) row into a relational database table. Under this approach, SQL can be used for retrieving location information. However, this approach is very inefficient when it comes to answering some specific classes of queries at the database level. In particular, location at a particular point in time cannot be retrieved unless location was recorded for this point in time in the database. Asking for the location of a mobile device at some point in time in the future cannot be answered at all. Asking for all mobile devices that are within a particular geographic region during a specific time window during the past several days requires a rather complex SQL query.

A large body of research exists that studies the problem of how to efficiently represent location information on top of a relational database. Most of this research focuses on the modeling and indexing structures that could be used for speeding up location-based queries. Some of this research extends the SQL language with trajectory management operations and semantics.

For context-aware services that track user appointments using calendar entries, observed routing patterns, or explicitly entered trip information, the following questions need to be answered based on recently collected location data:

1. Is the user headed towards a particular calendar appointment?
2. Is the user going to be late for the next calendar appointment?
3. Are there any external events that may influence whether the user will be able to reach a destination on time (e.g., weather, traffic)?

Using trajectory-related techniques only might not be sufficient for answering the first question with enough confidence. The reason for this is that there may be multiple routes available to reach the destination associated with a calendar appointment. A prototype service that addresses the above question is described in [FAL08].

5.4.2 Accessing Location Information

Location-based services depend on the most recent location of a mobile device in order to be able to provide location-specific services (e.g., location-based reminders and traffic alerts). As we discussed in Chapter 3, the most recent location of a device can be computed using one or more of the available positioning technologies in several ways. Depending on the available positioning technology options, LBS requirements, and mobile device capabilities, LBS software designers and architects need to determine how mobile device location becomes accessible to the LBS service.

Here, two main architectural alternatives are available: *push* and *pull*. Under the push approach, location information is sent by an application running on the mobile device or some other network-based component to the LBS service on a periodic basis. Under the pull approach, the LBS service retrieves location information from the device or some other network-based component on demand (note that the user does not see nor need to see the differences between these alternatives).

Depending on LBS functionality, the amount of location data available and processing requirements for such data may vary considerably. For example, let us consider a telematics service that enforces a virtual geographic fence around vehicles (referred to as electronic fence). The service triggers some form of an alert or takes some action when a vehicle crosses a particular geographic boundary. Assuming that the number of vehicles that may be monitored by such a service is in the millions, then the service would have to solve the problem of how to efficiently process location information retrieved from vehicles and check whether any geographic boundary violations occurred.

Clearly, one would have to parallelize the above computations to a high degree in order to support a large number of vehicles that have electronic fence enabled. This could be achieved in several ways. For example, one could split processing based on geographic rectangles and have one service instance responsible for each such area. When an electronic fence is set up for a vehicle, the service instances responsible for covering the geographic region specified in the electronic fence will be notified to start monitoring vehicle location events and performing fence violation computations.

As with any application that handles user data, security and privacy are two very important issues that need to be addressed in the context of LBSs. Obviously, being able to know where a mobile device is located at a particular point in time makes a lot of people uncomfortable. This is because, in most cases, a device can be associated with an individual either directly or indirectly. Letting users control when location information is collected, and when and how such information is being accessed and shared with other users and applications is the preferred approach. This is the approach followed by existing location aggregators, such as the Yahoo's Fire Eagle.[4]

However, for some location-based mobile applications the user may not have the option to opt-out. For example, a car rental company may enforce tracking location of its fleet of rental cars in order to enforce contractual terms with respect to where a rental car can go and cannot go. Here, a user renting a car will most probably not have the option to disable location tracking.

5.4.3 Partitioning

Given the potential huge volume of location information, how can one store such information in a way that provides scalable access? An obvious solution to this problem is to partition the entire geographic area covered by the offered service into several, non-overlapping, regions using simple geometric shapes, such as rectangles. By doing so, a specific location can be easily mapped to one of these regions and, therefore, location-based queries can be carried out in parallel for each region. When a particular region becomes overloaded due to either the number of location events associated with it or the number of location-based queries, this region can be further partitioned in smaller regions, each with its own dedicated processing and storage resources. Similarly, when adjacent regions experience light load, they could be merged in order to consolidate storage and processing resources.

The above approach works well for most cases. However, it presents some challenges when one wants to be able to track a mobile device that travels across different regions. A solution to this problem is to maintain a mapping between mobile devices and regions they cross. The information in such mapping facility will change slowly for most mobile devices and, hence, the management and maintenance overhead will be relative low. Furthermore, recent mapping information could be replicated across regions for more efficient accesses to such information.

5.4.4 Replication

Mobile services depend on information and content. Parts of the information and content are stored in relational databases. Some other parts are stored with file systems. Yet some other parts may be transient and exist only in memory. Typically, content created using computer and mobile phone applications or captured using a combination of

[4]http://fireeagle.yahoo.net

hardware and software (e.g., video and audio clips) is stored in files. Metadata describing content attributes, associations, classification hierarchies, and access restrictions are stored in database tables.

Independent of the storage medium used for storing information and content, the information and content must be highly available and readily available to geographically dispersed/mobile users. Replication is the technique used for achieving this. However, the replication of portions of the available contents and its associated meta-data presents many challenges. While various techniques could be used for supporting selective replication, such as a combination of database snapshot replication and file system replication, these techniques are not integrated and, typically, depend on "coupled" application, database, and file system designs.

Recent work in the areas of content delivery networks (CDN) and peer-to-peer networks (P2P) addresses replication [AKA] [BART] [JUN] [QLV]. In these systems, replication is used for improving response times and addressing fault tolerance requirements. Web pages and images are typically replicated among Web or application servers in order to improve response times in a CDN environment. However, this work does not address integrated database and file system replication.

In many cases, replication is enabled after a service has been operational for some time. This is a common theme occurring at start-up companies that are trying to offer new services in order to capture user attention. In many cases, user attention is extremely high and the service experiences a lot of traffic. In such situations, several challenges need to be addressed in order for replication to work. The following list provides some examples:

- Many system entities are assigned numerical keys that are not globally unique. To preserve uniqueness, new primary keys must be generated for database entries that are replicated.
- To facilitate direct access to content stored in file systems, relative path information may be stored in the database. When content is replicated, relative path information may need to be rewritten.
- Some content may reference other content. Therefore, replication needs to address referential integrity constraints.

In addition to the above issues, replication may have to be applied to only some of the available content. Content that is short-lived, propagated to available sites via external mechanisms, or not accessed that often many not require replication. Furthermore, available capacity at a particular replica and infrastructure costs need to be taken into consideration.

There are several replication approaches available. Typically, they belong to two categories: asynchronous and synchronous. Asynchronous replication maintains a queue of changes made to the underlying content and propagates such changes on a periodic basis. Synchronous replication propagates the operations or modified data to remote replicas as part of the original operation. There are advantages and disadvantages to both of these approaches. Depending on the content access characteristics, one or both of these approaches could be followed [PAN].

5.5 SUMMARY

In this chapter, we addressed always-on services and discussed two main properties of such services, namely availability and scalability. We discussed several design aspects of highly available and scalable architectures by providing concrete service examples from the area of telematics and intelligent telecommunications networks. We also covered some of the business aspects of offering always-on services, including infrastructure costs, and partnerships. Finally, we went over data management issues related to dynamic data associated with location-based services.

In the next chapter, we will continue our discussion on always-on services by focusing on the notion of "always working" services. These services tend to be long-running mobile services that not only do work on the user's behalf but take advantage of local resources in an opportunistic fashion.

QUESTIONS

1. How would you define availability of a service when the service depends on a telecommunications network that is provided by a third party?
2. How would you troubleshoot service issues from a user's standpoint when several parties are involved in offering the service?
3. How would you plan for scalability and address fast user growth? Perhaps lessons learned from rapid growth in telecom network subscribers are applicable here.

6

UNDERPINNINGS OF MOBILE OPPORTUNISM

> "The most profound technologies are those that disappear. They weave themselves into the fabric of everyday life until they are indistinguishable from it."
>
> —Mark Weiser

Today's mix of technologies, services, and information form a complex brew. More and more, new innovations in technologies allow richer services; at the same time, there must be a tradeoff, as defined by the "value landscape,"[1] between offering every conceivable feature and the simplicity and usability of the service or device. This chapter looks at the notion of opportunism, how to design and support it, and the ramifications on both client and server.

6.1 MANAGING COMPLEXITY AND OPPORTUNITIES

As we know (also see previous chapters in this book), mobile devices are computationally powerful and contain multiple network access technologies including cellular, Wi-Fi, Bluetooth, and IP. Built-in and add-on sensors such as gyroscopes, compasses, glucose monitors, and vital signs sensors are prevalent. Mobile device Operating Systems and middleware have also increased in capabilities, openness, and programmability, for example, see Windows Mobile 6, Linux and Android phones, and MIDP,

[1] See discussions in previous chapters.

The Fabric of Mobile Services: Software Paradigms and Business Demands, by Shoshana Loeb, Benjamin Falchuk, and Euthimios Panagos

respectively. Mobile device hardware and peripherals – such as keyboards, screens, and other I/O devices – have improved dramatically in recent years and device positioning and location based services have become mass-market. Each mobile device then, whether it has telephony features or not, can be considered a rich, locatable, mobile computing platform. With this robust platform comes great potential – potential for great new services but also for great confusion if the services are not well designed. With mobile services, both time and place become first class concerns and enablers. Both time and place contribute to the complexity of the life of today's typical "power user," someone characterized as being often on-the-move, traveling within and beyond his home city, involved with multiple "connected" activities on a personal level (such as blogging and social networks), having multiple identities, and having complex needs and desires with respect to commerce and scheduling. These users have complex personal and work schedules, consisting of meetings and tasks, and a rich set of associations with mobile, e-commerce, and information services. The user may have eBay auctions on the go, Amazon marketplace offerings, Flickr photo-streams, Twitter posts, and blogs; the former condition may imply that the user has several calendars (some shared, some personal) that describe all her commitments and many to-do lists that describe tasks for which she should find time.

It is generally agreed that we will have arrived at true "pervasive computing" when the "power user" described above can walk down a street and have his handheld (or wearable) device act as an intelligent personal assistant as it receives, generates, filters, and otherwise manages "opportunities" whose completion would increase the user's quality of life (Fig. 6-1). The user's devices on the wireless "personal

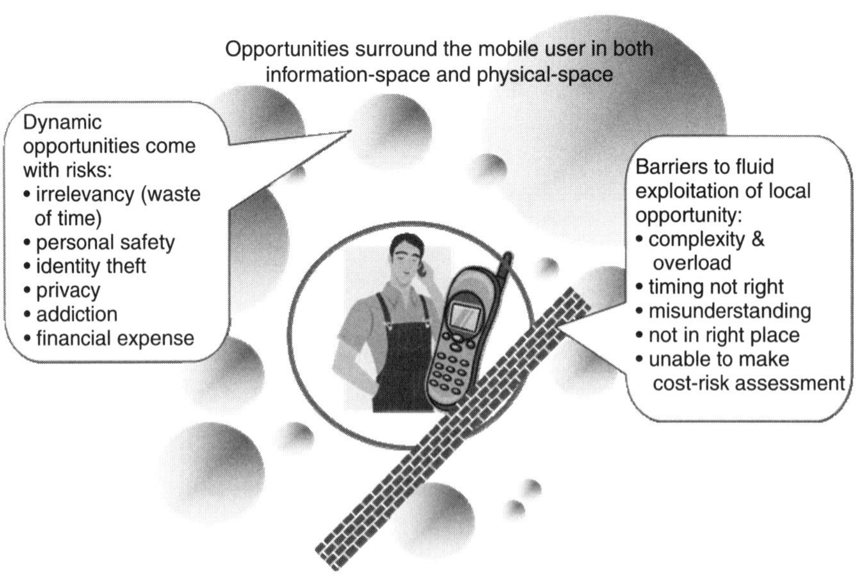

Figure 6-1: The mobile user is surrounded by opportunities of different types. Risks and barriers are also present.

area network"[2] (see Chapter 3) may suggest to the user to stop and pickup bagels as he passes a bagel shop, to go for a walk in a nearby park to get in her obligatory daily exercise, or to meet up with a person in a coffee shop because that person knows someone who is selling a motorcycle that the user wants. As shown in Figure 6-1, a mobile user can be thought of as being surrounded by ever-changing opportunities that are "close" to him in both the information space (e.g., the meta-details of the opportunity) and the physical space. If there are too many of these opportunities to self-manage, brokering (sometimes called "match-making") can occur through an intermediary system. Successfully brokering opportunities can be complex and requires some of the following: 1) getting the timing of the notification right, 2) supporting on-the-fly cost-risk assessment using either the rules of traditional economics or, as we mentioned in previous chapters, rules derived from "behavioral economics," 3) getting the notification details across unambiguously, and 4) minimizing information overload.

6.1.1 Opportunism

Despite the potential of mobile devices to provide value-added information and computing power in conjunction with application and network-resident servers, today's walled-garden telecom services and locked-down devices rarely allow users to experience the wide range of possibilities that their mobile phone *actually* supports though this is beginning to change with today's programmable phones. Location-Based Services (LBS) offer interesting information and navigation possibilities [FAL4] but are typically software "islands" that work only with proprietary phones, equipment, and situations. Because LBS services are largely built and offered in stove-piped form to specific network operators and handsets, users cannot simply make any application they want to be location-aware, and the locked binary versions of the services on the handset cannot be mashed up.[3] Mobile users would like to have the best possible services and information at whatever location they happen to be at in a particular moment, as long as privacy concerns and other preferences are respected. These services may be ones that were easily anticipated or ones that were unexpected and provide value by fitting somewhere into the "long tail" of the user's preferences. Ultimately, opportunism should result in a benefit to some actor in the ecosystem, for example, the user, the network operator, the local service provider, a local business, etc. The latest mobile devices already bring great potential for human-to-human social benefits, given that previously unrelated mobile users can be linked together with a communications channel no matter where they are, based on any number of parameters (e.g., consider a social network that links people together

[2]The "personal area network" is the advanced concept of the local short-range wireless network that can interconnect any device either on the user's body or in the immediate surroundings.

[3]While standards exist at the physical, link, and network levels, information services running on proprietary operating systems do not interwork well. This is usually because the service data architecture is closed, with no APIs.

who are similar in spirit but also physically close to each other). On a darker side, the possibility for digital stalking, deception, fraud, and other sorts of exploitation does exist [FAL4]. Before 2.5G networks and location-based services, while the mobile device could certainly receive text messages, there was little in the culture or business ecosystem that resulted in value adding opportunistic services. Even though your mobile device was always on your belt buckle, it was not playing a role in your moment-to-moment interactions with the world, apart from receiving incoming phone calls, text messages, and perhaps e-mails. That has changed, however, as 2.5G and 3G devices are enjoying fast growing market penetration and the use of text messaging, picture-taking, and GPS supported services is on the dramatic incline, fine-grained location-based context-aware services (e.g., startups like RememberTheMilk and OhDontForget) are starting to see the light.

Context-aware systems are generally those in which any information related to the user, device, object, or location is exploited in order to adapt the system's behavior. These systems generally sense and/or communicate with the physical environment through one or more mobile devices carried by the user and also those embedded in the environment. [BALD] surveys a broad history of context-aware systems, including vintage but influential ones such as *ActiveBadge* (in which a user's position is transmitted to the system via infrared and the system routes subsequent phone calls to devices nearby). Through the 2000s context-aware computing has become a broad research area and has dove-tailed with the area known as *pervasive computing* [WEI2], which some say that Weiser kick-started with his 1991 declaration that, "the most profound technologies are those that disappear. They weave themselves into the fabric of everyday life until they are indistinguishable from it." [WEI]. Location-based services can be thought of as a class of context-aware pervasive services, and, up until recently, have been seen mostly as data and information services. More recently though, these new services are concerned with linking people together and creating social networks, circles of trust, and collaboration.

Opportunism, as a concept within the broad research realms of both pervasive and context-aware systems, is yet a more recent phenomenon and is subtlety different than pervasiveness and context-awareness. It implies that the mobile computing system senses an opportunity to exploit a local resource and takes action in a way that was not necessarily completely defined or expected a priori. In other words, in opportunistic services either or both of the following are true:

- The resources encountered by the device are not entirely predictable in advance; no "script" for processing them in a particular occurrence exists because no such encounter can be guaranteed
- The actions taken by a device on a local resource work to the benefit of the device's end user or some other entity in the mobile ecosystem

By "guided" or "informed opportunism" we mean that the mobile device takes its local action based on some guiding principle seeded by a global authority or server – in other words, not completely based on knowledge gained "locally." The seeding may happen in bulk or asynchronously via bi- or uni-directional communications between the seeder (e.g., a server) and the device. Thus, in opportunistic mobile services one may see:

- Explicit directives from a server (or peer) to a client to support resources and knowledge discovered locally by the client
- More loose suggestions or advice from a server (or peer) to a client to support resources and knowledge discovered locally by the client
- Either of the above in isolation of locally discovered resources

StumbleUpon, *Reddit*, and other similar web sites can be considered opportunistic crowd-wisdom-based Web sites because they present information that can be both surprising and useful and reflect the aggregate views of a large user-base. This kind of opportunism can also already be seen in simple form in a few emerging commercial wireless services:

- Wi-Fi based geo-positioning – mobile device reports previously geo-located Wi-Fi beacons to a server as it sees them, is positioned through correlations with a database of hot-spot locations, and then receives location-based information from the server [RAP]
- Route planning – making use of services found along a route in optimal way, e.g., finding the best gas value on a navigated trip [HOR]
- Mobile dating – alerts (followed by text messaging) when compatible singles are "nearby" (see startups *Crush or Flush*, *MeetMoi*, *Zogo*, and so on)
- Network usage – advantageously choosing locally discovered transport networks [FAL3]

Opportunism generally exploits location and position (see Chapter 5), though it need not – for example, a system taking proactive actions based on changes in financial data could be unaware of user location and yet still valuable. In addition to informational resources, those in the physical space, such as kiosks with network and computing capabilities, network access points – both wired and wireless – and even the computing resources (laptops, handsets) of other users, might also be candidates for exploitation and usage. For example, assuming a suitable business agreement, while roaming around, one's handset might transparently use spare computing resources from a nearby user's laptop; or find files there that satisfy its requirements (such as operating system or middleware components). In order to distinguish the above sorts of computing services, the complexity level of the opportunism can be a differentiator. In Table 6-1 we illustrate the differences between some current and emerging systems in this class. Since the area is still in its infancy, we expect this list to evolve rapidly in the next few years as innovations and new services are introduced at a rapid pace but in tune with the ever-growing value landscape (see Chapter 1) and the vision of ubiquity.

In Figure 6-2 we depict the paradigm of opportunism in some detail, along with the other actors in these sorts of systems. In the figure, the term "agent" stands in for any software component that plays a specific functional role in the use case – whether or

Table 6-1: Types of opportunism and existing systems

System	Description	Level of Semantic Complexity of Opportunism
2G phones, no rich services	Older phones and services could essentially only take incoming calls and texts	None
2.5G SMS-based messaging	With wide-adoption, support, and popularity of SMS, subscription-based services sent text notifications upon dynamic events. For example, device gets a text message when their favorite NHL hockey team scores a goal, or when a stock price exceeds a threshold.	Very low to none (trigger-based sending of messages)
In-car navigation systems	Today's in-car navigation allows turn-by-turn directions as well as points-of-interest searches.	Very low – the information base is preloaded and queries are highly limited and structured a priori
Mobile dating and other one-off applications	Mobile dating systems have emerged that use the device's local sensing resources (e.g., Bluetooth, Wi-Fi, or server-based triangulation) to interact with other mobile users. When two devices are nearby and both device users are registered "daters" a text message can inform them that a romantic opportunity is possible at the moment. Very rich, but one-off – applications in research labs have appeared in which device context is exploited further.	Medium – a mobile "dater" has a fairly simplistic profile and the computational complexity of a "match" is not difficult
Mobile binding	Mobile media storage and playback devices such as the Zune player specifically encourage opportunistic sharing of media with other mobile users, but offer little semantic. Current research allows the incorporation of geospatial options into sharing decisions [FAL2]	Medium to high – the semantics of direction, velocity, and acceleration may come into play; in addition, the semantic of information sharing may be complex

(Continued)

Table 6-1: *Continued*

System	Description	Level of Semantic Complexity of Opportunism
Improved in-car navigation and mapping	Current research and commercial and academic sectors will soon result in better in-car mapping and route-planning by incorporating a) anticipated delays and other costs to the user, and b) deep-context aspects from the users personal information systems (e.g., scheduled appointments, preferences, etc.) [FAL5]	Medium to high (see discussion later in this chapter)

not the software displays true "agency" is the topic of a later subsection. The rest of this chapter describes many of the aspects shown in the figure in more detail.

6.1.2 Opportunism and Service Design

This section outlines key issues relating to opportunism and service design and planning. Some of these issues are of general concern, but all should be considered

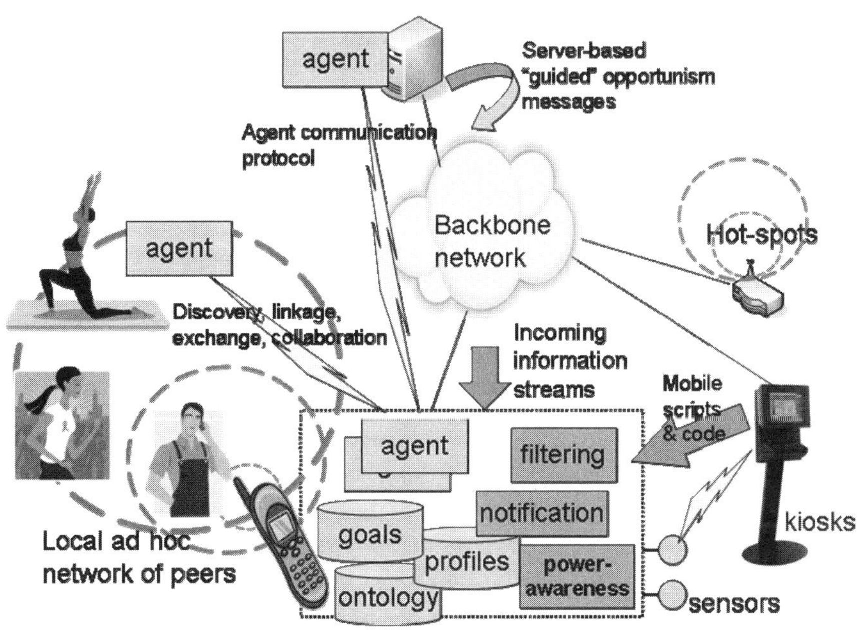

Figure 6-2: A graphical summary of some of the entities and systems involved in opportunistic wireless services.

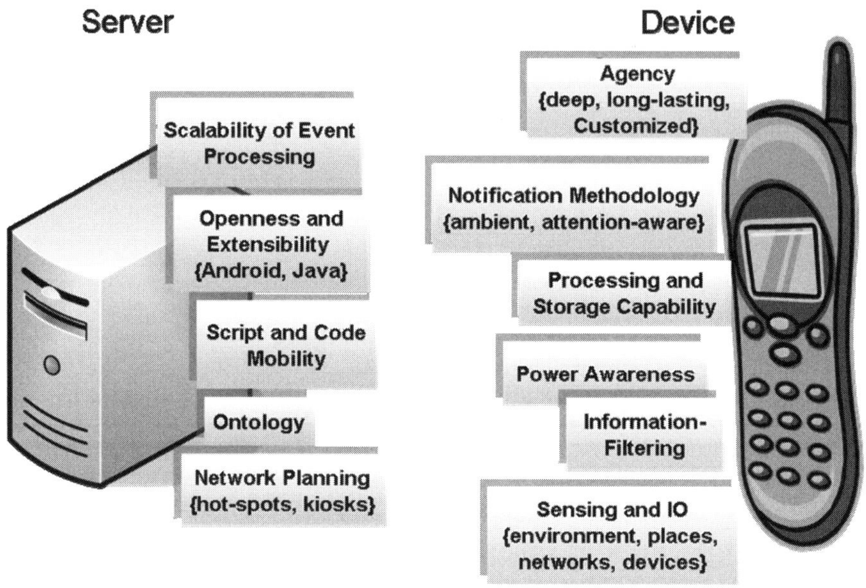

Figure 6-3: Division of labor in supporting opportunism and the computational issues for operators (servers) and users (devices).

when planning a proactive or opportunistic system. Subsequent sections describe particular technologies in more detail. Figure 6-3 presents some selected key issues and classifies them as either server or device-side concerns (with some simplifications for the purposes of this discussion) – device positioning is a separate concern and studied in detail in Chapter 3. In this section, we take server to mean a role in which data is stewarded (or can be accessed) and is queried by a client, and some global view of the system is possible. One way to think of the server is as a managed entity, owned by the Network Operator, or Service Provider, with a set of interfaces to access the data it may need to manage the opportunism of a set of mobile device clients. Nothing prevents the server from being comprised of many distributed components (possibly mobile themselves!). The client on the other hand is the equipment co-located with the mobile user and has both up and down links to the server. The client either polls the server or subscribes to notifications; in response the server sends guidance down to the client which, in turn, may allow the client to sense, connect to, and exploit local services, resources, and information.

Server-Side

Scalability of Event Processing On the server side, scalability, availability, and event processing are of key importance due to the scale of both the user base and the number of events generated per user at any given time (see Chapter 5 on the challenges of "always on"). Event-driven Architectures (EDA) (see Chapter 7) are highly relevant to Opportunistic computing. Why? Because efficiency and timeliness are extremely important (a suggestion given "too late" is "useless"). In EDA systems loosely coupled information sources such as mobile devices and remote computing systems transmit information towards information consuming systems (such as a server operated by a provider). The consuming system stores, filters, and reacts to the incoming events accordingly often by forwarding it to other information sinks that have expressed an interest in the content. These "interest subscriptions" are either: a) subject-based – the subscriber identifies a logical channel name (e.g., as in CORBA) or a topic (e.g., as in Java Message Service (JMS)), or b) content-based – in which the subscriber provides a set of parameter-values or patterns that should be met – these comprise the constraints on the desired information and can, ultimately, be highly complex. In these architectures, event dissemination data structures can be highly formal and structured (e.g., XML, schema), or informal (e.g., flat text – as in a stock price or news update). In general, all such structures can include:

- A combination of both subject and data information;
- Event name and parameter-value pairs or event-descriptors;
- Rich XML information.

Event dissemination-reaction may occur in one of several modes:

- Simple – a single received event triggers some action.
- Streamed – a stream of events arrive in succession and must be correlated and acted upon in some aggregate form. In this case an intermediary event filter will often perform aggregation and re-emit a new event for "simple" event consumers.
- Complex – a mix of event types arrive and various event management techniques are required.

Correlation of events to each other typically occurs on one or more of these bases:

- Temporal;
- Spatial;
- Casual (ad hoc).

Openness and Extensibility In opportunistic system design, openness and extensibility are issues applicable to both server and client. An open system, or open-source system, refers to one in which the software that built the system is available to anyone to inspect or change. Changes to the platform are documented and made available to the developer community for further action. Extensibility refers to a system whose functionality can be extended either in whole or in part. Extensibility mechanisms may include rewriting or creating new source code that builds the system, or plugging in some other technology that effectively changes the functionality. These concepts are exemplified in the Open Handset Alliance, "a group of more than 30 technology and mobile companies who have come together to accelerate innovation in mobile and offer consumers a richer, less expensive, and better mobile experience. Together we have developed Android^{TM}, the first complete, open, and free mobile platform." The decision on openness and extensibility can have a profound effect on the service's impact. Will third parties be able to see how this service is implemented? Will third parties be able to tweak functionality and add their own custom features by changing and rebuilding the software? What is the business ecosystem surrounding the opportunistic service? Will the service be portable across wireless operators or work solely in a walled garden? Open Handset Alliance promises a future in which services can be written by third party providers on both large and small scales. Notably, the J2ME and Symbian ecosystems *already* enable a more limited form of code portability and openness. J2ME especially, while not a top to bottom system architecture, has deep penetration and millions of developers and compatible devices in the large.

Script and Code Mobility Script and code portability and mobility are an interesting research topic. A typical mobile service written to conform to the Java MIDP specification has a deployment option called "over the air" (OTA) (other technologies have similar mechanisms). In OTA, a compressed, binary version of the Java program comprising the application is sent on demand over the wireless network to the mobile handset where it is uncompressed, installed, and started for the user. True "code mobility," on the other hand, is an advanced use case often found in tandem with systems exhibiting a degree of agency (more later on this topic). In a code mobility scenario, an executable program makes not just one hop from server to client, but possibly several hops after the initial one, travelling from platform to platform. At each platform the program is executed and delivers some value either to the local platform or to the owner on whose behalf the program executes. Such a program is sometimes called an "intelligent mobile agent" and has been the topic of research since the 1990s [MAE]. Intelligent, mobile software agents will almost surely play an interesting role in future opportunistic systems. One reason is that the development paradigm can employ abstract goals, plans, and use high-level speech-act based communication utterances, much like a human user would. Therefore, the paradigm can sometimes be a better mapping for long-running tasks that require planning and communications. In other words, the constructs used by the developer to develop the agent are similar to the cognitive ones used by human end users (see also Chapter 4). A hypothetical mobile agent-based opportunistic scenario follows: A mobile user sitting on a campus green wants to see a report of the social networking profiles *of the people sitting nearby (assuming their devices are running compliant platforms).* An agent is created, discovers the peers nearby and is transported, in sequence, to each peer device's execution platform. At each platform the agent interacts with a device-resident profile agent and engages in communication acts that result in information exchange. When the agent has finished it sends a summary of results to the originating mobile user. This sort of scenario has long been played out in the lab across controlled homogeneous platforms, but not in a consumer market. Other barriers to mobile agent development include securing both the computing platform and the user's privacy.[4]

Ontology and Semantics In order for the software entities in an opportunistic system to intercommunicate, understand the current context, and make a decision, they must have a shared semantic. The semantic of an operation or communicative action refers to the analysis and representation of the true "meaning" of the action, as opposed to its syntactic representation (a given meaning can be conveyed in different syntaxes). In turn, in some formal systems, this representation is encoded into a schema, ontology, or other canonical form that all participants understand to be the formal definition of semantic [TGR]. In some cases – such as by way of ontologies created using best-practice tools such as Web Ontology Language (OWL) – the formalism is machine-readable and can be processed by any software machine with the appropriate parsers and reasoners. Machine-readability results in the ability to create distributed intelligent components that communicate over a shared semantic, thereby greatly reducing the ambiguities and misunderstandings that are dangerously present in systems with no formal footing. It also allows agents that never planned on "meeting" each other to communicate opportunistically; in other words, the use of ontology greatly assists the development of systems in which it is may be hard or impossible to predetermine the entities that will be encountered. As a simple example, the mobile agent scenario given above requires that the device resident agents and the mobile agent have a shared understanding of what each request for information implies. That is, the mobile agent is asking the device agents for social network profiles – if the device agent misunderstands the syntax (e.g., the encoding) or the semantic (e.g., the meaning of the request) then the requester agent will surely fail to acquire the information it needs).

Planning Also on the server or infrastructure side, the service provider needs to plan how resources are distributed to support the opportunism use cases. Are the local resources managed by the operator or by another entity? Does the planner even know what the resources are in advance? Physical resources may include networked kiosks[5] placed in urban settings, or network hotspots. In some cases, other mobile users with compatible mobile devices are also considered resources (i.e., the Zune scenario), and indeed this is the peer-to-peer (P2P) model in which peering endpoints engage in a temporary information exchange session using the network. The P2P model can result in very beneficial services for users but tends to use the network as a "dumb" pipe, leaving the network operator out of the ecosystem. In general, opportunistic service creators will want to minimize their management overhead and, unless the physical resources are already an intrinsic part of their business

[4]The paradigm is seen mostly in military and industrial applications rather than consumer ones; the paradigm is emerging and useful; a good example of a mass market device with a much weaker sense of opportunism is the Microsoft Zune device; Zune is programmed to share its media (songs, playlists, etc.) with other Zune devices it detects nearby.

[5]Microsoft Surface technology (http://www.microsoft.com/surface/index.html) is a viable option for a kiosk operating system, providing multi-touch interaction on a large (up to 30 inch) tabletop.

model, will often want to build information-centric services that depend only on the presence of other users or on the usage of resources already implemented on the mobile handset (e.g., the Bluetooth network, the on-board camera, etc).

The above are only a sampling of the types of decisions and rationale that might go into planning the "server-side" of an opportunistic service. The advantages of the server-side include its global view of the network conditions (via Operations and Support Systems), and its practically unlimited storage potential and processing speeds, at least as compared to the mobile device.

Client-Side

Software Agents Figure 6-3 also presents some of the issues of importance on the client side of opportunistic services. We have already established above that in order for opportunistic services to run effectively they need to somehow "reach out" and communicate with the mobile user. The computing environment that runs on every smartphone as well as so-called *bodynets* or *piconets* (a local network that links the devices on and around the user) and wearable computers of the future are the platforms physically closest to the user. The mobile device may also be acting as a source of information or may be being asked to modify its internal state according to some directive from a server or peer. For all of these reasons, agent-based technologies are desirable. The role of the device-resident intelligent software agent might include protecting sensitive information, providing access to device data as per the owner's wishes, and determining which opportunities best fit with the user's personal goals – both explicitly and implicitly declared – of the user. With respect to goals:

- An explicitly communicated goal may be captured directly by user input via some interface on the device. Broad examples of these sorts of goals include: the entry of a new scheduled event (e.g., to call a friend at 13 : 00h), the specification of a particular information search criteria, or the creation of a "to-do" entry that defines a series of subtasks and due dates.
- An implicitly inferred goal is not entered directly by the user but is inferred by a software agent that is armed with any of: past history of user behavior, detailed information about user profiles and preferences. In order to parse and manage such information this agent must be able to mitigate the semantic representation(s) of this data and must implement an algorithm for determining goals from these "clues." As a simple example, an agent with access to past call history might detect a pattern (e.g., user calls his mother nightly around 8 pm) and opportunistically message or setup the call on evenings wherein the user forgets to do so.

It is useful to separate the functionalities of a) detecting and inferring the correct moment and information with which to take an opportunistic action, and b) the mechanism by which the user is notified about this information (if at all). The agent take care of part a) while using platform APIs to deliver knowledge and information to the user.

Figure 6-4: A notification example for Android. For the developer of this notification, a few simple invocations of the notification manager achieve the desired effect. (Source: http://code.google.com/android/images/anr.png)

Notification The emerging area of notification management is increasingly interesting as mobile devices have ever greater capacity to interact with the user through multimodal and graphical interfaces. While ambient systems such as a smart storefronts or retail spaces employ custom-built immovable equipment (e.g., LCD screens, kiosks, smart surfaces and fabrics, and so on), mobile notification systems rely on the device in the consumer's hand. Today's smartphones have a reasonable amount of screen space on which to render graphics, and can – in some cases – run sophisticated sensing and inferencing software. Notification on the handset is concerned with:

- Providing a service to notify the user of events (which can be used by other system services)
- Integration with system-level functions that allow notifications to be made on multiple modes (such as with graphics, sounds, vibrations, and so on)
- Weaving some sort of intelligence into notification either in isolation or in conjunction with other agents

System-level notification management, such as that found on the Android platform, is defined as a "Class to notify the user of events that happen . . . [and] how you tell the user that something has happened in the background." Notifications can be "a persistent icon that goes in the status bar . . . flashing LEDs, flashing the backlight, playing a sound, or vibrating."[6] Figure 6-4 shows a simple notification in Android, generated by an invocation of the manager by another running program. Though it is promising

[6]See Android documentation for details at: http://code.google.com/android/index.html

that notification appears as a key programmable part of Android, a forward-looking view for "intelligent notifications" would most likely entail adding mechanisms such as software agents (e.g., "the notification agent") to augment and implement Artificial Intelligence (AI) capabilities (see Chapter 1). In this case, we can envision other components on the device, such as "opportunity agents," hoping to notify the user of an opportunity make a request to the notification agent (including parameters of interest along with the request).

An emerging aspect of notifications is to implement highly ambient and nonintrusive methods; Weiser describes it as "calm computing" [WEI3] and we can be fairly certain that increasing focus in this area will yield further interesting approaches such as those that use:

- Silent or colorful notifications;
- Customized and personalized ring-tones for differing notification types;
- Auditory cues in a highly multimodal way (e.g., a "whisper" overlaid upon an existing conversation);
- Biophysical interaction in the cases where the devices are a part of a body-net;
- Attention and contextual awareness – such advanced scenarios would include the ability for the notification agent to take into account current location and activities relevant to the user and mitigate notifications to take best advantage of them; also, deriving the user's current attention (such as where the user is looking and what task is currently being focused on) could considerably improve notifications.

The Palm Pre, released in 2009, features an innovative notifications bar that enables elegant anytime notification presentation.

Processing and Storage Processing and storage are additional key factors in implementing opportunistic services. There is a clear need for highly sophisticated processing and storage on the server where the server stewards the data for a large number of customers simultaneously; on the mobile device side the processing and storage requirements are a function of how the opportunism is implemented. Server-based "guided" opportunism puts much of the onus on the server side, and very little on the device side. Device-side opportunism is not seen in the larger market today but does show up in academic and commercial research labs and is more and more feasible every day. A device-side implementation relies on locally executing long-running software agents, local knowledge or database, and algorithms that implement the opportunism. In such a scenario, the device might need to be able to process many pieces of independent information and fuse them together without the help of components or data on a central server. Additionally, it may have to run complex algorithms and store a relatively large amount of information (such as profile, statistical data, and service-specific data) to deliver value to the user.

Purely as an illustrative example, consider a mobile application that advises the on-the-go user on how to manage her personal wellness throughout the day. The handset's sensors stream biometric readings such as heart-rate and blood pressure to be read by a software component. Other sensors on the device might detect the presence of compatible handsets nearby or merchants and storefronts. Using a database of metadata relating locations to services and relating activates to population density, the system might compute a plan for the user: 1) suggest the user stop at a park and get some exercise, 2) suggest other "compatible" people nearby with whom a game of pickup Frisbee may be interesting, 3) instant message people nearby with an invitation to a game, 4) cross-reference local businesses selling healthy products and navigate the user to them. We can see that in order to run completely on the device side, this advanced use case requires quite a large database of information (including maps, places, and other people in the system), services, and heuristics. Finally, the need for processing and storage is even more acute in emerging peer-to-peer (P2P) use cases in which mobile devices play both client and server roles for distributed content. The P2P paradigm, which proved incredibly compelling and controversial for wired PCs, is the subject of a great deal of research and commercial attention. While there are still unanswered questions, there is no doubt that massively distributed storage and routing of information through P2P and social-network-based associations is sure to be an active research area for several years. In mobile P2P, each device will need to play a role as an intermediary for data storage, lookup, and processing.

Power Awareness An overarching concern of all services, but especially of opportunistic ones, is mobile device power consumption and power awareness. In wireless sensor networks (in which routes between nodes are formed in so-called ad hoc fashion) power awareness is important. Every message sent from the node and every computation made on the node drains the battery; when enough nodes fail the entire network may cease to be useful as a propagation network. In mass market mobile computing, there are few to no power-aware computing issues that bubble up to the consumer, except perhaps the display of the battery level! Beneath the covers, however, there is a good deal of research on how chipsets can most efficiently use power in order to maximize lifetime, but of course this is not directly accessible to consumers (though they see the effects via the battery icon). For example, an ARM11 process has several distinct power modes including: 1) "run mode," in which all processor functionality is available, 2) standby mode, in which the device is still powered up but most of the system clocks are powered down, and 3) dormant mode, in which the core powers down but the caches and TCM remain up.

For developers who use low-level languages such as machine code, power-awareness can be developed "into" the software by using registers and hardware in an intelligent way. For most developers, who use higher-level languages, there is little instrumented power-awareness possible. Battery level instrumentation (i.e., the availability of APIs to enable programs to read and understand current battery level), however, is widely supported on mobile OSs and development kits. Figure 6-5 shows custom warnings that can be instrumented in custom applications to warn users when levels are low.[7]

[7]Source: http://www.devx.com/wireless/Article/34696/1954

Figure 6-5: Using sophisticated mobile OSs such as Windows Mobile 6, developers can instrument notifications triggered by device battery levels. Opportunistic intelligent device applications anticipate and notify in a more systematic fashion.

The reality of consumer electronics is that low and dying batteries can cause negative session events, such as dropped calls or poor quality. In this regard, opportunistic battery-aware technologies should be of interest to carriers, as customers sometimes do not distinguish between call drops that occur because of network conditions versus those that occur due to local battery issues (thus the blame is placed unjustifiably on the carrier's service). Poor session quality can also be caused by:

- Weak signal strength as sensed by the phone. This, in turn, can be due to geography, climate, motion, etc.
- Network congestion that prevents the phone from getting the appropriate resources needed for the establishment and continuity of a session.
- Phone-related errors such as misconfigurations, low battery, etc.

In [RAV] the authors propose that opportunistic battery anticipation technology on board a mobile device should always strive to answer the following question: Will the phone battery last until the next charging opportunity is encountered? One way to answer this question is for the device to predict call duration times and correlate them to anticipated battery levels. When insufficient battery levels appear, a notification is sent to the user. Another way to provide power awareness is through "battery presence." For example: while A and B are sharing a voice session, B's battery level goes below a "dangerous" threshold. Agents on the device anticipate that the current session will be interrupted by a battery shortage very soon and ensure that at this moment A's view (and B's view) of B's battery level is updated (on an icon on A's screen, or via some other notification), thereby giving A full awareness of B's situation. The result is that A and B avoid a "surprise" session disruption, and even if it cannot be avoided (B has no AC adapter at the moment) A and B can verbally make a "backup plan" such as, "if our session drops, let's talk again at about 3 pm." In summary, power and battery awareness and opportunism are a key part of emerging opportunistic services. Taking advantage of power surpluses and scaling back in the presence of power shortages is an intuitive and, one has to think, useful paradigm.

Information Filtering In opportunistic systems, useful intermittent notifications might be streamed to a device from two source types: servers on the network and the local device software. In both cases the senders are hoping to steal a little bit of the user's attention with the notification (e.g., a popup window, an SMS message, a reminder, etc.); however, the user has only a certain amount of attention that can be borrowed at any given time (see "attention economy"). Now, if there are a large number of servers and peers competing for a user's attention, there is the potential for "information overload" and this is where information filtering technology comes into play. In some senses it will be similar to e-mail spam filtering in which many unknown senders send unsolicited time-wasting e-mails to one's inbox. However, in an opportunistic system, unless the endpoint address is compromised in some way, filtering is more likely to be necessary within and between the opportunistic services that the user has knowingly subscribed to. When we think of an opportunistic system we think of functional component such as:

- Application – the principle logic;
- Message marshalling – part of the application that initiates and requests a notification or message to be sent to an endpoint;
- Low-level message sending and receiving – operating system or low-level API through which messages are encoded and sent over one or more physical links to the destination.

We should also think of the information flow as two-way: Servers and peers send information towards a given mobile user, and the mobile device senses the environment, does some computation, and sends information towards a server or peer. Interesting services might be built from devices reporting about:

- type of applications currently open on the device;
- type of activity currently being performed on the device;
- biometric (e.g., human vital signs) and biological (e.g., air quality, radiation) sensor information;
- set of local networks currently "seen" by the device (e.g., SSIDs of Wi-Fi networks);

- set of local peer devices currently seen by the device (e.g., user-IDs, Instant Messaging handles);
- information on the photographs stored locally on the device (e.g., taken by the device's camera).

In response to these reports a server may respond in kind with a directive relating to:

- a network to use (e.g., to receive discounted service bills);
- a number to call (e.g., to receive further information);
- a location to go to (e.g., to receive services).

Table 6-2 summarizes interesting opportunistic applications and exactly how information filtering could come into play.

Information filters may have local or global scope – local filters may be specific to updates from certain applications, while global ones apply to all in- and outbound updates. Information can be filtered based on the content of the information, the situation and context of the user, time of day, and many other parameters and combinations of parameters. Content-based filters detect tokens and/or patterns in the

Table 6-2: Scenarios

Domain	Description and What It Filters
Location identification via Wi-Fi beacon reporting	Device sends updates to server on the Wi-Fi SSIDs that it currently "sees." System filters out SSIDs that don't belong to the set of those that are seeded in the system, owned by the provider, etc.
Time-based updates	Device sends constant updates of the device's current processor load. The filters constrain updates such that this information is sent "at most 5 times per minute," or, "not until after 6 pm," etc.
Health reporting	Device wants to report subscriber's vital signs over communications network. The System suppresses updates if signs are in-range or unchanged from previous updates.
Application and "future intent" reporting	Device wants to inform server which applications are currently active on subscriber's device (or that will be active soon). System filters out those applications not of interest.
Peer reporting/social networking	Device wants to update server on local in-range peers detected via layer 2 beacons (e.g., Bluetooth). System filters out certain peers or peer-types as possible.
Communication analysis reporting	Device updates the server on the phones communication activities (e.g., outgoing calls, incoming calls) but the system filters out those numbers that are of no interest to the server (e.g., 1-800 numbers)

formed update message and trigger actions that, in turn, reduce the size of the update message (e.g., filtering something out). Situation filters are triggered when particular conditions surrounding (not within) the content are present. Filters that are able to filter out content based on some saved past content are useful (e.g., allow the update to pass only if it's different from the last one). Time-based filters allow the current time, time-stamps, time-thresholds, etc. to play a role in filtering. Time-based filters can reduce communications used by updates by completely blocking out updates in certain cases. For example, the following types of blocking, or throttling, time-based filters could be described and activated:

- Update rate constraining – described in plain English as, "UA should not send more than *n* updates through the next *t* time-units."
- Update blocking – described in plain English as, "UA should not update server again before absolute time *t*."
- Filters of either (or both) types can be combined to make more powerful and intelligent update filtering scenarios (e.g., filter on both time and content).

Thus we see that filtering of information is a key technology that will go hand-in-hand with opportunistic systems, keeping them scalable, manageable, and resource-efficient.

6.2 TECHNOLOGIES SUPPORTING OPPORTUNISM

This section outlines some of the interesting technologies that support aspects of opportunism.

6.2.1 Agent-Based Systems

Since high-profile research initiatives in the mid-1990s – notably at MIT and IBM – there has been a renewed interest in making software systems more intelligent and responsive. The media attention and promise surrounding agent-based systems peaked during the 1990s and, while we hear less about this underlying technology today, past efforts are now visible in key niche areas. Manufacturing, industrial scheduling and resource management, and military applications are among the areas that make use of agent-based technologies and standards [WOO]. Web phenomena such as the natural language questioning possible at Ask.com, the collaborative filtering of Amazon.com, the music recommendation at Pandora.com, and the pro-activity of notification services such as iwantsandy.com, although not purely agent systems, validate the appeal of the effects of agent-based systems (i.e., collaborative filtering techniques can allow the site developers to make visitors feel like an intelligent agent is watching them and suggesting things that "make sense"). Agents can be defined as, "computational systems that inhabit some complex dynamic environment, sense and act autonomously in this environment, and by doing so realize a set of goals or tasks for which they are designed" [MAE]. Agent-Based

Systems (ABS) – usually implemented in object-oriented fashion – display a subset of the following:

- Consist of loosely coupled and/or distributed agents;
- Communication is via message passing (at a high semantic level) of an asynchronous nature;
- Do not typically have a single centralized administrative or execution controller;
- Agents are transient and play various, changing roles;
- May be under-specified or fail gracefully ("confusion" doesn't mean "failure");
- Capture, at a fundamentally low level, abstract notions of intent and desired effect;
- May display emergent behavior not explicitly programmed or expected a priori.

What does this mean for mobile services? Complex mobile services can be built using software agents as the underlying enabling architecture. As a case in point, long-running, personalized processes are important to the evolution of mobile services since notification, monitoring and reminder services all work best when run continually as a background process (any downtime in operations may mean missed opportunities or events). In other words, these services run on the device even when not visible on the interface and when other applications are in the foreground. The Open Handset Alliance (OHA) Android platform allows for such long-running services to be spawned; a typical Android example is to use a long-running service to handle the playback of a song while other applications run. Support for background processes on the iPhone is much more contentious.

The ability to exchange knowledge with other agents is a key part of *agency*. The ability to communicate with other agents not known to be present at design-time, and to do so in an asynchronous message-passing paradigm is powerful but challenging. To achieve such communications agents typically need to "speak the same language." The FIPA.org standards include the definition of a knowledge wrapper format for speech-acts that allow individual pieces of information to be intercommunicated. There are three levels to intercommunication:

1. A shared understanding of a structured format that holds knowledge;
2. A shared understanding of the usage of descriptors that describe the content and its semantics;
3. The embedded knowledge (or information, or payload).

Regarding the specification:

> "[it] does not mandate any one formalism for representing message content. Agents themselves must arrange to be able to interpret any given message content correctly. . . . In particular, a set of standard communicative acts and their meanings is defined. It may be noted, however, that there is a trade-off between the power and specificity of the acts. Notionally, a large number of very specific act types, which convey nuances

> of meaning, can be considered equivalent to a smaller number of more general ones, but they place different representational and implementation constraints on the agents. The goals of the set of acts presented here are (i) to cover, overall, a wide range of communication situations, (ii) not to overtax the design of simpler agents intended to fulfill a specific, well defined purpose, and (iii) to minimize redundancy and ambiguity, to facilitate the agent to choose which communicative act to employ." (Source: http://www.fipa.org)

Therefore, as a service developer, the reasons to use agents in the service design might include the desirable trait that the agent can interwork with other agents on the device, share information, and even collaborate to achieve a shared goal (provided other agents adhere to the same standards). This allows for more reactive, long-running, and adaptive mobile services rather than static and unchanging ones. Examples of hypothetical forward-looking agent-based services running on a handset might include:

- Monitor system levels and interact with the user: An agent runs continually on the device's OS, monitors battery levels, and profiles application usages. When the agent detects battery is approaching low values it considers past usage of applications and anticipates when the battery may run out. It may also have an understanding of the user's schedule via the device's calendar, so putting these all together, and inferring a battery loss in 10 minutes, it communicates this to the user through a pop-up window.
- Interact silently with other agents: A personal agent runs continually on the device and stewards profile information about the user's preferred musical genres and statistics regarding music files on the device (e.g., how many, etc.); intermittently other agents (representing other users that are nearby) are transported to the device over the air and begin execution. The new agent intercommunicates with the device agent, gathers answers about musical preferences, migrates off of the device, and then reports its findings to the nearby user. An option then allows this user to text-message with the other in the case that musical interests are highly similar.

Figure 6-6 illustrates agent communications at a high level. Inside the logic of the agent, the developers encode the agent's goals (e.g., what it would like to complete) and the agent's intent (all actions it has committed to making). The agent's main execution thread is tailored to continually plan and execute actions until the goal is reached.

In summary, in this section we have discussed the importance of agent-based systems and possible architectures of future mobile services. Today's mobile Web and relatively simple device-resident applications have not, for the most part, required or called-out for the agent paradigm. Imminent mobile services, characterized by needing asynchronous communications with other agents, long-running goal-directed behaviors, and opportunistic actions, will likely see the same benefits from agent-based systems that military and industrial applications already see today with agent

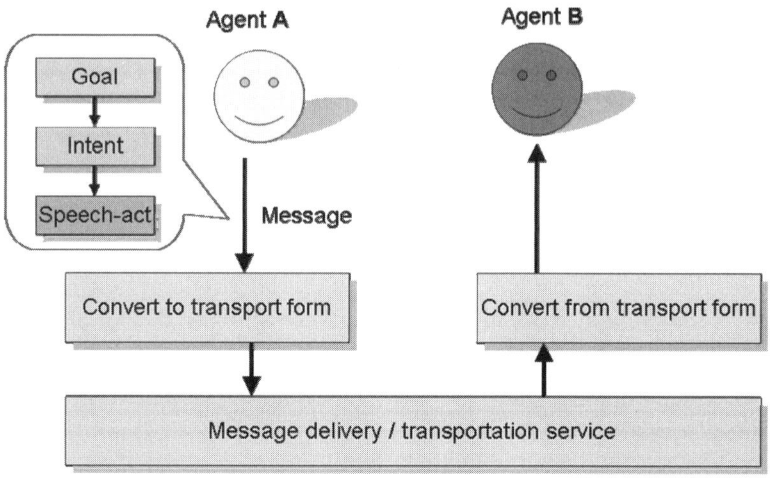

Figure 6-6: Agents on the mobile handset intercommunicate in order to deliver a service to the user.

systems. We are already seeing the demand for these sorts of services in more and more sophisticated mobile reminder systems and personal assistants. Some recent evidence that agent technologies may again be gaining steam and momentum: for two years running a dissertation on the topic of intelligent agents has won the British Computer Society Distinguished Dissertation Competition. According to noted agent researcher Nick Jennings, "We are now moving towards practical devices that support the effective (agent-based) co-ordination and formation of teams of first responders in major disaster response scenarios."

6.2.2 Semantic Web and Ontology

The semantic Web is the next generation Web of information (sometimes referred to as "Web 3.0") whose (invisible but) structured markup reveals much more machine-readable meaning about the content than does today's HTML. Tim Berners-Lee has declared that, "properly designed, the Semantic Web can assist the evolution of human knowledge as a whole." Semantics of intercommunications and data interpretation are key factors in successful implementation of agent-based and opportunistic systems. Misunderstanding between such systems may result in "soft errors" in which the communication does not fail, but the action's effects are unwanted and unanticipated due to error. In opportunistic systems, soft errors may result in the wrong notification sent to a user or to poor timing. In Web systems, a simple example of a semantic misunderstanding is when a Web search engine misinterprets a search for "Jaguar" as a search for the animal of the same name, rather than a search for the vehicle as might have been intended.

The technologies leading the pack for Web-scale semantics include Resource Description Format (RDF) and Web Ontology Language (OWL). These technologies allow the creation of controlled vocabularies called *ontologies*. An ontology specifies a domain of discourse, and represents concepts in a hierarchy tree and the relations between them. It can also represent constraints and axioms. RDF and OWL, therefore, give humans a syntax with which to encode information with respect to the ontology and give computer programs common machine-readable codes from which they can derive a deeper meaning than simply the codes' syntactic structure. Indeed there are several realms in which semantic Web is of rising importance:

- Mobile search – the ability to search for information from a mobile phone (often for services "near" the searcher) is a major beneficiary of semantic Web technologies (e.g., Caboodlenetworks.com); because navigating search results on a phone is usually difficult and awkward, it can be a competitive advantage to use ontologies and rich search semantics to deliver higher quality results that relate personal context and profiles to search results.
- Mobile peer-to-peer information sharing – using the co-ordination and reasoning capabilities of semantic middleware to optimize the exchange of information between mobile hosts (e.g., may involve understanding geospatial aspects of movement, etc.).
- Intelligent recommendation systems – finding the right information for the user based on deep understanding of preferences and information meta-structure.
- Social network analysis – understanding how people are both directly and indirectly "connected."

Current barriers to semantic Web supported mobile services include:

- On the client side, a lack of semantic Web APIs for small footprint devices.
- On the server side, a lack of business-class tools that can be used to model and manage ontologies, and tailor them to mobile services, equipment and paradigms.

The tools market is changing though, thanks in part to the continued full-court press from Tim Berners Lee, who drives the research and gives thoughtful and motivational presentations on the topic, to R&D such as that at IBM and Microsoft, and an industry that is starting to see value in offering semantic Web tools (e.g., Semantics Works, Oracle, etc.).

In contrast to ontologies, a folksonomy is a shared intelligence, created from the bottom up by a group of users. Typically, the lexicon is developed in order to improve search, or *findability* [MOR]. Flickr, del.icio.us, Technorati, and a host of other Web sites allow users to contribute both to the folksonomy (e.g., by creating new tags) and to the indexing of information (e.g., by attaching a tag to an article). A given site's tag-cloud is an interactive graphical representation that depicts the popularity of particular terms in the folksonomy (usually by using a larger or emphasized font for those terms). Because they do not encode equivalence or hierarchy, folksonomies are less machine readable than formal ontologies in which, for example, terms are encoded with self-describing hierarchy, equivalence, rules, and properties (such as transitivity, and inverse relationships). Many researchers believe that folksonomies simply do not scale as well as hierarchical systems. As a simple example, looking at a tag-cloud on Flickr one might notice the tags "happy" and "sad," through which one will find joyful and remorseful images, respectively. However, nothing in the tagging infrastructure can indicate if these terms are consistent with the rest of the folksonomy, whereas a formal ontology could encode the relationship between the two terms, and thus allow richer browsing and search use cases.

6.3 DESIGN OF OPPORTUNISTIC SYSTEMS - EXAMPLES

This section describes a small set of interesting and representative opportunistic systems, and illustrates the various technical and design choices that are necessary.

6.3.1 Opportunistic Network Utilization for Mobile Devices

We have been involved in the design and implementation of a pilot service named *Argus* that is highly illustrative of opportunistic network usage [FAL3]. In this system, a server provides opportunistic guidance to mobile handsets (MH) to indicate which of several locally visible networks the device should use for communications sessions at any given instant. The system takes its actions transparently to the user and exploits the global view afforded by the server, thereby distinguishing it from other interesting efforts such as T-Mobile *HotSpot @ Home*. The system benefits both end consumers (e.g., their devices attach to the best network automatically) and network operators (e.g., by controlling guidance carefully, network loading and blocking rates can be decreased). The main actors (also shown in Fig. 6-7) in this scenario are:

- MH and Messaging Agent (installed on the MH Operating System) – the MH must be, in this case, inherently multi-modal, supporting at least one additional network interface technology beyond cellular; e.g., Wi-Fi or WiMAX or both. Bluetooth is on the radar for body-nets and so on, but is not relevant here.
- Policy Server (in service provider's infrastructure) and other infrastructure servers and gateways to both enable and manage intercommunication with MHs.
- Transport channel for bearing protocol messages.
- Local exploitable resources: In this case, they are Wi-Fi hotspots and cellular service (co-managed by the same service provider or by business partners[8]).

Of primary interest here is how voice-calls originating from the MH use and affect the local networks. The technology is useful because an Operator can use it to help

[8]If the MH is to be guided intelligently by a 3rd party then it must have had some insight into the underlying attributes of the resources it recommends. In the case of voice network resources, metrics like total capacity, background traffic or anticipated call-rate would be considered in the 3rd party's recommendation-formulation.

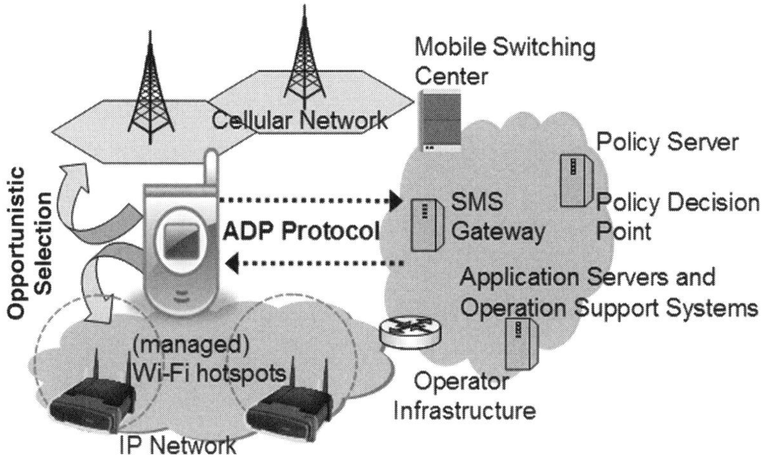

Figure 6-7: High-level architecture of the system. Guided by a server, the mobile device opportunistically uses one of the available networks for voice and application sessions.

reduce call blocking experienced by customer by influencing the device to use VoIP or cellular calls depending on network conditions (e.g., background traffic) at the location of the user.

In Argus the server and the MH communicate via an information protocol. The protocol – named ADP – is a step towards true device peering and autonomous information sharing and strives to offer a systematic way for mobile peers to transmit information and information updates, directives (e.g., commands), and queries. With ADP the server and the agent on the MH communicate – the server can tell the agent something, ask it something, or command it to take some action. In addition, we imagine the agent as wanting to convey sensed information to the server (there are some similarities to FIPA's Agent Communication Language here). Thus, the messages we use in the opportunistic network usage scenario are: REGISTER, UPDATE, REQUEST_PREF_LIST, OVERRIDE, PREF_LIST, REQUEST_CONNECTION, ACK, and ERROR. The first four of these are used from the mobile device to the policy server; the next two are used from the policy server to the device. ACK and ERROR can be used either way.

An UPDATE message semantic is used to exchange network resource context. The ADP Client sends UPDATE to the Server following a change in the state of any of its layer 2/3 information such as networks to which the device has a layer 2/3 attachment and "visible" network beacons. The UPDATE message serves to keep the Server up-to-date with the mobile device's context. UPDATE may be acknowledged by the Server with the transmission of a new preference list. UPDATE is similar to REGISTER except that may or may not be acknowledged by a PREF_LIST. In fact, a field in the UPDATE message can be used to indicate whether such a response is required. Other variations of the message allow the ADP Client to form more compact messages by indicating incremental changes to previously transmitted information. As an example, an ADP Client in the presence

Table 6-3: The preference list details

Probability	Sequence	Net. Type	Net. ID	Key
0.6	1	WLAN	fappoi2	*******
0.4	2	Cellular	telcowireless	–

of two Wi-Fi hotspots with ACME cellular service would encode that information into an UPDATE message for the server. The Server may respond with the appropriate PREF_LIST message.

The Preference List (PREF_LIST) message is sent by the Server to inform the ADP Client of a new set of preferences, or "guidance"; in this context it encodes preferences for local network usage (optionally associating networks to specific device services). PREF_LIST is not acknowledged and it should be possible to broadcast a PREF_LIST to many users in parallel via a broadcast channel. Table 6-3, though simplified, shows some of the data in a preference list.

The Preference List is an essential way for the Server to convey probabilistic network preferences to ADP Clients. In the table, the Server guides the client to select the Wi-Fi network with probability 0.6 and the cellular with 0.4. If the hotspot "fappoi2" is Wired Equivalent Privacy (WEP) protected then the key is also piggybacked. Figure 6-8 shows a high-level use case of the system in action (see [FAL3]). After a

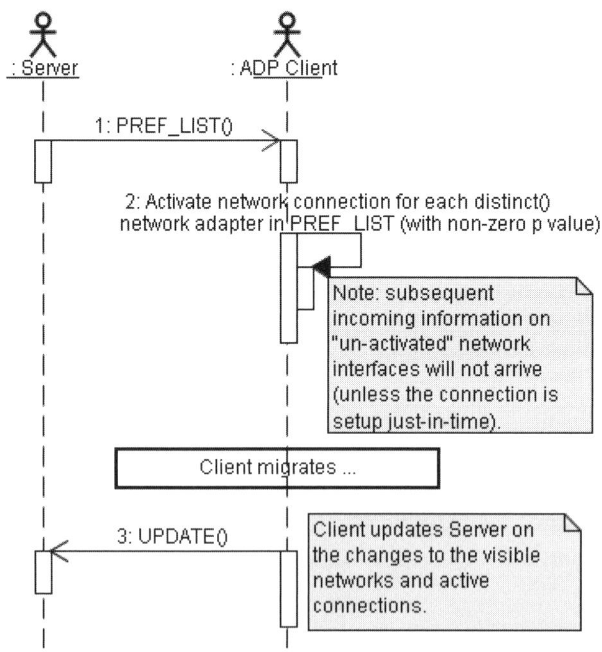

Figure 6-8: ADP Client receives PREF_LIST but later migrates; this warrants a new UPDATE message capturing networks at the new locale.

server sends the mobile client a preference list of networks, the software on the device takes several actions including activating network connections on the networks indicated in the incoming preference list. When the client moves to a new geographic region the handset detect new networks (such as new Wi-Fi beacons and new cell towers) and reports them up to the server. The server uses its global view of network conditions and the handset's current location to decide upon the preference list to be sent to the user (and the cycle repeats).

In summary, the Argus system and its protocols are examples of how the users' unpredictable movements through heterogeneous and multi-network coverage areas can be made more systematic through opportunistic software components that, due to their global view of the network conditions, can give the mobile handset guidance on how to use local network resources to the best result.

6.3.2 Opportunistic Navigation and Reminder Systems

Opportunities exist all around us. They impact various aspects of our lives and, when missed, we may pay a price. When we are driving, in particular, we cover a great deal of geography and often cut across urban settings. The current state of the art in in-car GPS navigation systems offers accessibility to novice users, low-costs, and reliability. Typical in-car Personal Navigation Devices (PND), however, are data islands. A more recent trend is for INS has been to have IP connectivity via satellite or cellular links built into the device but this is still an exception (see Chapter 8 regarding current trends in this area). There is still very little momentum for linking PND with external, open, personal information systems relevant to a given single user.

In [FAL5] the authors describe a "deep" context-aware computing in which many aspects of a given user's context are linked to PND through a high-speed, centralized, event engine and database. Incoming events are read and stored at a high-speed and a context structure for each user is constantly updated. Service logic – such as triggers based on locations and time, rules, and individual preferences – is incorporated into an application server. Capturing and understanding user context is key to being able to offer personalized and relevant services. Here, user context is considered to be an intersection of facts, artifacts, events, and circumstances associated with a user. The following are some of the most important components of user context that must be captured and managed on a streaming basis:

- Personal information (e.g., interests)
- Social information (e.g., contacts)
- Subscriptions (e.g., traffic alerts, feeds)
- Device capabilities (e.g., vehicular)
- Presence and availability
- Privacy and access control
- Schedule (e.g., calendar entries)
- Present status (e.g., location, time, activity)

While some of the above components may not experience rapid state changes (e.g., privacy and access control), others may generate quite a large number of events (e.g., when users change locations, when calendar events are due, prescriptions expire, when the automobile speed is over a threshold, etc.). There is also composite context and a large number of rules to deal with. In addition, the user will demand very short reaction-response times (e.g., suggestions to buy products after the user is too far from the grocery store are useless).

In addition to handing events originated by sensors and other sources (e.g., calendar appointments, presence information, GPS receiver), successful personalized telematics services such as this one need to be able to anticipate user needs based on context elements that may not be directly associated with specific time or location, such as general to-do list entries. Here, an opportunistic approach is required to address user needs in a proactive manner and to enhance the value of the service. The main benefits to the user of such a system could include:

- Location reminders – a mobile user's "to-do" list includes planning a party. The system "understands" that this entails buying supplies, thus, as the user drives "close" to a grocery store, the system reminds the user to buy supplies.
- Context migration – as a mobile user enters or leaves another compatible computing area (such as his car or his home) a subset of past context is migrated from the mobile device into the new environment, allowing richer inference to be made between it and the handset.
- Intelligent suggestions – a mobile user moves between regions while in a communications session with a peer. The new region has an a priori known communication inhibitor (e.g., is mountainous) and the system, reacting to the location-change, anticipates session interruption and provides a tangible "warning" to the mobile device indicating "session disruption possible – make alternate plans." In this scenario our Event ending receives "region-change" event via network probes or other third party positioning services (perhaps from the handset itself).

Figure 6-9 shows some aspects of the system in question. As the user travels through an urban area the back-end system constantly consolidates opportunistic suggestions that address explicit or implicitly inferred user context (e.g., to-do tasks, appointments, etc.). These opportunities are presented (in this prototype) to the user on a map; the system uses text analysis in order to suggest a nearby CVS Pharmacy when one of the user's to-do's for the day was to "order my prescription." The system also shows an explanation of why an opportunity is being presented and provides directions and other details. Through use of filtering techniques, the system does not overwhelm the user with either irrelevant suggestions or too many options.

The deep context system presented above is, at the time of writing, beyond the state of the art. However, this sort of innovative navigation system is starting to appear in limited form in PNDs and also in in-car systems. Mobile smartphones are beginning to play the PND role as well; thus the players in this value-chain include: PND manufacturers (Garmin, TomTom, etc.), cell phone manufacturers (Apple, LG,

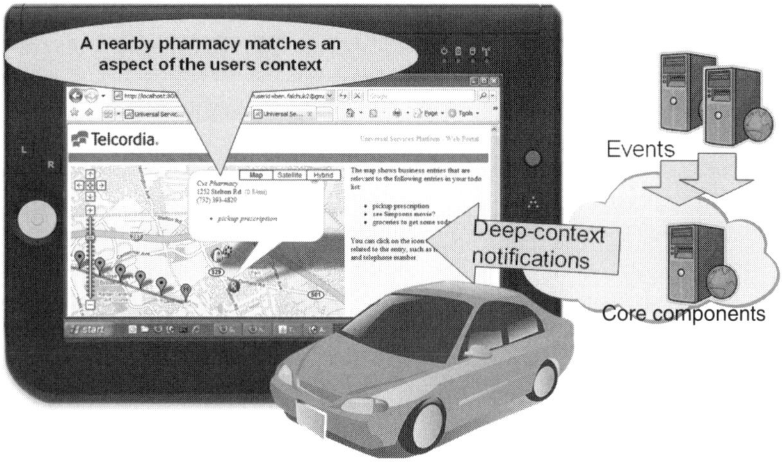

Figure 6-9: The big picture – a system for generating intelligent suggestions to mobile and driving users.

Samsung, etc.), application developers, content providers (map providers, advertisements), and many others. The prospect of presenting dynamic context-aware advertisements during navigation is also a hot topic and is on the fingertips of researchers around the world.

6.3.3 Mobile Dating Systems

Mobile dating systems are already in the market and some comprise a particular type of limited opportunity. In general, the first generation of online dating allowed single users to 1) complete a personal profile, 2) post the profile online, 3) be found by another registered user through the portal, and 4) message (voice and instant message) with the user for the purposes of setting up a date. As an example, Match.com has more than 15 million registered subscribers with profiles. A first push into mobile space came when several of the online Web-only players enabled their subscribers to both check their dating portal messages and to text/instant message with other singles using their mobile smartphones. However, a truly more opportunistic mobile dating system would go beyond the above scenario and new generations of such systems are now available. In these systems a positioning technology reports the user's location to the dating system; in response, the system uses correlation techniques to determine if any other daters match the following criteria: 1) are nearby, 2) are open to ad hoc communications via the cellular phone, 3) have compatible profiles. If all conditions are met then the portal may inform one of the mobile users that a compatible single is "nearby," as well as kick-off other service logic to enable a meet-up between the two matching users (e.g., one user might get a text message with the content, "Joe – a registered compatible single user is nearby – she's open to communications – if you'd like to text her *click here*"). The positioning technology (see Chapter 3) is the new enabler here (step 1), while the other proprietary algorithms of the dating system need not change (steps 2 and 3). A main reason that dating portals are interested in providing this convenience factor is that they can charge a premium for the luxury of such location-based notifications. A Juniper Research report estimates that mobile dating revenues may exceed the $1 billion mark by 2010. Finally, the bandwidth of 3G and heterogeneous 4G networks makes IPTV and video messaging possible. Not unexpectedly, an ecosystem of video dating messaging is currently emerging in which the mobile handset is used to record and view video messages that help daters decide on compatibility issues (e.g., enable love at first sight, so to speak, using high-speed networks). So opportunistic mobile dating systems are a new and exciting niche application area that may eventually deliver the following sorts of services:

- Notify a mobile dater when compatible singles are nearby the users current location.
- Provide information layers that overlay city maps that display where single people (with certain attributes) congregate.
- Send a video to a mobile dater via the cellular network when requested by the user or when a backend system anticipates that such a video is opportune.

It goes without saying that participants opting in to dating services (mobile or otherwise) voluntarily give up some of their privacy in exchange for the possibility of a "match." The location enablers are not adding this as a new compromise but *are* adding the user's current location as a new commodity. With any such personal data, there are great risks if the data falls into the wrong hands and great possibility for annoyance if the location provider sells out one's location to a spammer. Note that "dating services" are an example of a more generic category of services where people or services are matched based on profile and availability (including location). As a result, the infrastructure for mobile dating and other mobile services which were developed for the "consumer" market can also be deployed with some modifications the enterprise market, where the details of the tasks are different, but the structure is the same.

6.3.4 Mobile Localized Information Sharing

Opportunistic peer-to-peer information sharing means that information is not routed through a centralized server but is relayed between two more mobile devices using their own networking capabilities. In such a use case the following steps may occur:

- Device A detects device B by sensing a communication beacon from B on a well-known channel.
- Device A (or B) uses current conditions to decide if such a link is desirable (this includes an analysis of the services desired by A and offered by B.

- A link level connection is made between A and B.
- Information is exchanged between A and B using a well-understood (or at least jointly understood) semantic and syntax.

Link level and service discovery protocols such as Universal Plug and Play, Service Location Protocol (SLP), Jini, and Salutation, employ registries and string-matching to match client's needs with registered services. Much past work has been done to enrich the semantics of service matching by, for example, expanding registry/search semantics to allow for rich service querying, for example, both Bluetooth and OSGi discovery protocols have been embellished with additional semantics beyond interface names. Why would we want this? Because, in order for a software component (e.g., on the mobile handset) to be able to connect to and work with one of several available services, it needs to understand what the service offers and how to communicate with it. For sophisticated information services (e.g., mobile peer-to-peer protocols), the answers to these questions are intricate and require highly descriptive syntaxes.

In the desktop world, information sharing without a central repository is referred to as peer-to-peer (P2P) computing [THE]. P2P, while possible in the mobile device world, is not widely executed yet on the mobile platforms, mostly due to capabilities of the current generation of Operating Systems and the problematic issue of ever-changing IP addresses. The Microsoft Zune is a mobile media player with a built-in Wi-Fi interface. Zune's capabilities explicitly call-out for peer-to-peer media exchange. Assuming the appropriate settings are made, when one Zune owner encounters another, an option exists for either of the two users to selectively share playlists and media with the newly localized peer. The shared media is typically locked under digital rights management restrictions but nonetheless, the paradigm is powerful *and* opportunistic.

More recently, proposals such as those in [FAL2] propose that peer-to-peer bindings – as in the Zune use case – be mitigated automatically by a computational "broker" that takes geospatial factors into account. These factors might include:

- GPS (or other location) coordinates
- Instantaneous or average speed, direction
- Other service attributes, such as sizes of media that are ready to be exchanged

By taking geospatial factors into account the binding software components can truly be opportunistically intelligent:

- It may choose not to exchange with a candidate peer because they will not be in range of each other for very long (this can be inferred from current speeds and possibly bearings).
- It may choose to take current physical location into account as a mitigating factor for information exchange, that is, in some cases it *may* be desirable to disable information sharing in particular geographic regions (e.g., in schools).

6.4 RELEVANT STANDARDS

Currently, no standards directly address opportunism per se. Wireless standards do define the "home server" concept (e.g., Home Location Register – see Chapter 4), in which much of the subscriber's service and personal information is stored. While this centralized approach works well for many mobile services it fails to meet some of the requirements for ad hoc mobile P2P opportunism since it sits in the network (not locally) and is under the administrative control of a single party.

A communication protocol for knowledge interchange between software agents has been proposed and is used in a variety of commercial and government systems. The governing body is called Foundation for Intelligent Physical Agents and the Agent Communication Language comprises a shared semantic of the pragmatics of knowledge interchange and inter-agent communications[9]. The semantics of both brief one-off exchanges and longer-running conversational interchanges are defined. For example, two (compliant) agents running on a mobile handset may engage in a "negotiate" conversation in which one of the agents proposes something, the other agent counter-proposes, and so on until one of the agents either accepts or aborts (see the specification for definitive details). This sort of conversation between intelligent agents on the device would theoretically enable sophisticated mobile services in which agents dynamically and interactively achieve their goals through intercommunication. For example, one agent – sent from another mobile user – may negotiate with a handset-resident one in order to "purchase" some device-resident media (such as photos, videos, or music). FIPA's work is mature and is based upon solid theory and fundamentals. Agent Communication Languages have been used in collaborative agent systems to great effect in manufacturing, industry, information gathering and fusion, and in many other realms.

Today, service discovery protocols have many forms. Some are standardized, while some are simply de facto industry standards. They include:

- Bluetooth Discovery layer
- Salutation Consortium
- Service Layer Protocol
- Universal Plug-n-play (UPnP)
- OSGi Service Platform
- Jini

Service discovery protocols are intended to enable mobile devices to find services of interest in a network environment that is either unfamiliar to them or is changing dynamically. Services may be found on both the wired infrastructure platforms or on the mobile ones. This is directly applicable to mobile P2P services since before two "strangers" can intercommunicate, they must have a link connection between

[9]See http://www.fipa.org

them and must have a shared understanding of how to ask for and use services on the other's platform. The need for service discovery is even more overt in the Zune use case described above, in which mobile ad hoc device pairings are explicitly *sought out* in order to initiate media sharing.

6.5 SUMMARY

There are very exciting prospects for a new generation of mobile services that we will truly be able to call opportunistic in nature. These services will see embodiments of highly personalized, long-running, goal-directed agents that work on behalf of a user or organization. They will use and exploit local resources and peers in ways that are not hard-wired but that generally adhere to the semantics and rules of interactions. They will allow users to find and use services near where they currently are and will allow mobile users to "fire and forget" requests down to the agent on their handset, and simply be notified upon completion. Once tasked, agents on the device may make a number of interactions with other agents that reside on the device, may use the Internet to communicate with agents on remote servers, or may use link-level technologies such as Bluetooth and Wi-Fi to communicate directly with peer agents on other mobile phones and platforms nearby.

The paradigm is compelling because, for on-the-go users, keeping track of long-term and standing tasks as well as managing the connections and interactions with local services and agents is an onerous task that is best left to the machinery of software. Making these sorts of long-running personalized "butlers" a reality will require adding "intelligence" somewhere in the mix – probably into the mobile device agents as well as into the servers and systems that support the interactions. It will also require and depend upon advances towards the vision of ubiquitous computing in which devices interconnect and communicate with little or no attentional awareness required from the end user. In this sense, one day we may simply assume that while we are travelling about our mobile device is always "working on something" for us and waiting for the best moment to provide us with useful information.

In the next chapter we will examine the underlying software design paradigms and patterns that, just like the example we just discussed of the software agent paradigm, are instrumental to the development of the evolvable software fabric that creates sustainable mobile services infrastructures and applications.

QUESTIONS

1. What might be valued more by a user: opportunistic assistance with the minutiae of daily life (e.g., buying mix, mailing letters) or with the bigger "goal based" life tasks (e.g., learn to read Spanish). Can mobile services help in these cases?

2. What kind of "learning" apporaches might be needed so that opportunistic services can effectively make good suggestions to their users? Should we expect these suggestions to get better and better as the agents understand the user's daily actions or is life too complex?

3. Will kinds of dialogues might be necessary between human user and the opportunistic agent in order to describe the tasks that need to be done? Will graphical interfaces be used? Natural language? Others?

7

DESIGN PATTERNS FOR MOBILE SERVICES

> "Computational processes are abstract beings that inhabit computers. As they evolve, processes manipulate other abstract things called data. The evolution of a process is directed by a pattern of rules called a program. People create programs to direct processes. In effect, we conjure the spirits of the computer with our spells."
>
> —H. Abelson and G.J. Sussman
> *Structure and Interpretation of Computer Programs*
> MIT Press, 1985

The fast adoption of mobile devices, from consumer cellular phones to business productivity smart devices, has significantly expanded the capabilities available to these devices and increased the number of mobile applications and services available in the marketplace. The usability, scalability, and general value of these mobile services are affected by the overall service architecture, selection of software components, and implementation approach. For example, when service logic executes at an application provider's back-end systems and service interactions pass through a wireless operator's middleware components, the integration between the middleware and back-end software may be critical to the smooth operation of the service. The same holds for mobile and Web 2.0 applications that depend on a combination of JavaScript and AJAX (poor use of the technologies can result in inefficiencies and errors).

Gamma et al. [GAM] point out, "One thing expert designers know *not* to do is solve every problem from first principles. Rather, they reuse solutions that have worked for them in the past. When they find a good solution, they use it repeatedly. Such

The Fabric of Mobile Services: Software Paradigms and Business Demands, by Shoshana Loeb, Benjamin Falchuk, and Euthimios Panagos

experience is what makes them experts." Mobile service development is now at the point where it can exploit not only traditional and relatively mature design patterns but also benefit from the highly active and communal groups of mobile developers who work with open and emerging mobile platforms and technologies such as Java ME, iPhone, and Android, as well as service platforms such as Yahoo!.

In this chapter, we will address the technological challenges associated with the deployment and operation of successful mobile applications and services from the point of view of software design patterns. In particular, we will discuss several well-known design patterns and analyze their applicability and usefulness in the context of mobile applications and services. We will cover architectural patterns, user interface design, distribution and resource management, multi modality, and event-driven architectures. Finally, we will present a use case of a multi-modal content management service that utilizes design patterns discussed in this chapter.

7.1 BACKGROUND

In the 1977 seminal book *A Pattern Language: Towns, Buildings, Construction*, architect Christopher Alexander and his co-authors describe methods for constructing practical, safe, and attractive designs and provide a large number of patterns that could be used for different design scales [ALE]. Each of these patterns provides a description of a problem that occurs repeatedly in our environment and a description of a solution to this problem that could be used repeatedly.

Similar to construction patterns, software design patterns describe solutions to common problems encountered in software application design. Instead of solving these problems anew for each new software application, it is desirable to re-use existing, well documented, and generalized solutions to these problems. Design patterns simplify software design and help with code reuse and overall system architecture. What is important about design patterns is that novice software designers and architects can benefit from successful and time-tested techniques.

Software design patterns started gaining momentum after the publication of the book *Design Patterns: Elements of Reusable Object-Oriented Software*, by the famous gang of four in 1994 [GAM]. In this book, Gamma et al. describe several patterns used by experienced object-oriented designers for solving specific recurring problems. For each such pattern they provide a name, describe the problem that can be solved by the pattern (i.e., when to apply the pattern), outline the components of the solution, and discuss the outcomes of the solution and any trade-offs that may be involved. While design patterns do not dictate the particular approach followed by a specific software architecture or implementation, they do address problems encountered in user interface design, efficient algorithms, and scalable architecture.

7.2 ARCHITECTURAL PATTERNS

Architectural patterns describe the overall organization of a software system. In particular, they describe the various system components, the responsibilities of these components, and how they interact with each other. Architectural patterns cover a wide spectrum of system properties, from performance to high availability. It is for this reason that such patterns are of interest to software architects of mobile applications and services.

7.2.1 Model-View-Controller

The Model-View-Controller (MVC) pattern has gained a lot of attention recently due to its prevalent usage in Web site development. Typically, almost every interactive Web application contains a mixture of business logic, data access, user actions, and various ways in which to display data. For example, consider a simple Web application that provides movie search capabilities. A user searches for information about specific movies by entering the title, actor name, theme, or some other criteria in a simple HTML form. Depending on the specified criteria, some kind of a repository is accessed in order to retrieve matching entries and, perhaps, additional information associated with each matched entry, such as links to short video trailers. The results are then formatted and shown to the user on a Web page.

In the above rather simple service, movie information is most likely stored in a relational database and is accessed when a user performs a search. Based on the search criteria chosen by the user, the service needs to determine which information to fetch from the database (controller). Once this is done, the information is retrieved from the database (model), and it is formatted in a way that will make it convenient for the user to access this information (view).

Figure 7-1 illustrates two different ways in which the MVC pattern could be applied. Under the first approach, the controller is responsible for notifying the view (step 2a) after the user takes an action (step 1). The view then interacts with the model for the purpose of retrieving updated state information and providing feedback to the user (step 3). Under the second approach, the model notifies the view (step 2b) when a state change takes place. The notification may include the actual state change or just a change indication. In both approaches, the view would need to register to receive notifications with either the controller or the model. Such registrations could be *coupled* or *decoupled*.

Under the coupled approach, a view must interact directly with the controller or state and indicate interest in receiving state change notifications via some form of callbacks. Under the decoupled approach, a view registers with an event notification service to receive state change events published by the controller or state. Selecting which approach to follow depends on several factors, such as the overall service architecture (centralized versus distributed), business logic, and state change paths. The most critical factor is whether state can be updated either by multiple controllers in an independent manner or via direct interactions with the data repository, such as database updates by a DBA. If multiple paths can update the model, then the preferred MVC implementation approach is to have the model notify the views about state changes.

An important characteristic of the MVC pattern is that the same model can support multiple controllers and views. This is essential for mobile applications and services that support multi-modality. Here, a different view will most likely be used for voice

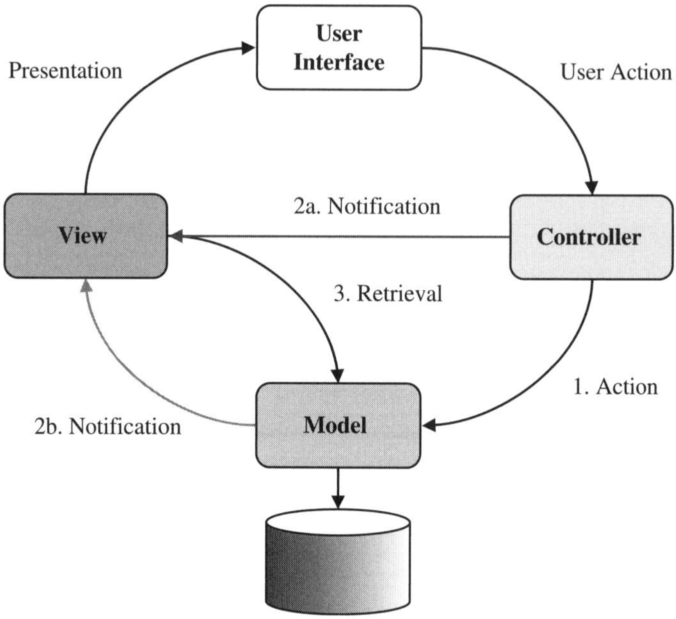

Figure 7-1: Model-View-Controller pattern.

interactions via a voice markup language (e.g., VoiceXML), and a different view will be used for Web-based interactions. Either the same controller or different controllers could drive these views. For example, imagine a multi-modal navigation service where the user asks the service for specific information via voice input (e.g., "next gas station") and the navigation device displays the retrieved information on its screen as an overlay on top of a map.

Some modern mobile software development kits do support (but do not require the use of) MVC for mobile development at a fundamentally deep level (e.g., assisting the developer at each part, making the MVC pattern overt). For example, Qualcomm's Brew UI Framework is one such tool. Other development platforms such as Java ME also enable the use of the MVC pattern, which is often used to perform actions and processes around an interactive interface on the mobile's screen.

7.2.2 Client-Server

In the beginning, mainframes ruled the IT world. Batch processing was the way to accomplish tasks, and applications were written in a monolithic format. Time-sharing minicomputers accessible by dumb terminals followed mainframes. Even then, the number of terminals were limited and placed in fixed locations. Users without their own dedicated terminals would have to wait in line for getting access to one of the available terminals. Similar to mainframes, the main computer performed all work, and terminals would just provide a simple input and output interface. Applications would still be written in a monolithic format and handle everything, from business logic, to data access, to user interfaces. After minicomputers, stand-alone Personal Computers (PCs) started to appear. Then, client-server arrived in the 1980s.

The client-server design pattern offers a message-based communication paradigm between two components, one acting as the client (requesting actions to be taken) and the other acting as the server (carrying out client requests). Client-server interactions may be either synchronous (the client issues a request and waits until the server responds), or asynchronous (the client issues a request and continues with its execution). The client and server components may be part of the same application (e.g., different execution threads) or correspond to two distinct entities running on the same hardware box or on different hardware boxes. Furthermore, a component may act as both a client and server and, thus, support peer-to-peer and multi-tier software architectures.

Splitting an application into client and server components offers several distinct advantages compared to monolithic applications. In particular:

- Resource consuming computations can be encapsulated into a server component that executes on powerful hardware, while the client component can run on devices with limited processing resources (e.g., mobile phones). By doing so, the scalability of the application can be greatly enhanced by either making more resources available to the server hardware or utilizing several server components.
- Access to shared resources (e.g., file system, communications channels, GPS receiver) can be carried out by a server component that is responsible for coordinating client operations and ensuring the integrity and consistency of shared resources. In the case of a mobile phone running multiple applications that require data connectivity, a server resource management component can coordinate accesses to the data channel and, in addition, enforce policies related to access priorities across data and voice sessions.[1]
- Modifications to application processing logic can be isolated in the server component only, without requiring any client side modifications.
- Proprietary algorithms, architectures, and information can be protected behind a server component, allowing companies to safeguard their technological edge.

Today, many of the Internet protocols are based on the client-server model. For example, HTPP is a client-server application where the client, typically a Web browser, requests information from a server using a well-defined message exchange mechanism. On the mobile device, many common applications are architected following the client-server pattern. For example, recall the A-GPS positioning technology discussion. The mobile device (client) asks the network-based A-GPS server for assistance in computing its approximate location. Because mobile client-server applications require message exchanges over a network, optimizing such exchanges

[1]Today's 2.5G and some 3G mobile communications protocols do not support simultaneous voice and data sessions.

is extremely important for mobile services due to device capabilities, potential bandwidth limitations, and monetary costs that may be associated with such messages.

Existing mobile devices support a number of different communication channels, such as voice, SMS, cellular data, Bluetooth, and Wi-Fi. Since the bandwidth, reliability, and timeliness properties of these channels vary (e.g., SMS supports up to 160 character messages and offers best effort delivery), selecting which communication channel or channels to use for message exchanges has a direct impact on both the functionality and quality of the offered service. Furthermore, the client-to-server and server-to-client communications channels may be asymmetric. This is because in most of today's cellular networks a server cannot establish an IP connection with a mobile device. Rather, the server might send an SMS message to "wake up" the mobile device, asking it to establish a network connection to the server.

Because of the above practical issues associated with mobile device-to-server communications, the implementation of the client-server design pattern in a mobile context requires careful consideration and in-depth analysis of the characteristics of the devices, networks, and target environments where the application is deployed.

7.2.3 Minimalistic Core

Many of us have heard anecdotal stories about software applications that offered a large number of features and functionality but took too long to deliver, ran over budget, and were hard to enhance and integrate with other applications. In most of these cases, the applications ended up not being adopted by users because of their complicated, and sometimes confusing, interfaces, or because they did not really solve the problem they were supposed to solve, or because they were not able to adopt to technological changes and keep up with competition and user demands.

How could one avoid the above situation while building applications and services for a mobile world? The KISS ("Keep It Simple, Stupid") cliché comes to mind. By avoiding unnecessary complexity and encapsulating the core application functionality into a small number of components, one could argue that the application will both solve the problem it is supposed to solve and be able to support enhanced functionality with minimal effort by delegating such functionality to add-on or external components.

The minimalistic core pattern (also referred to as *microkernel* in some cases) addresses application design complexity by separating the core functionality from the extended functionality. Extended functionality can be integrated with core functionality in a number of ways, such as via well-defined APIs, message passing mechanisms, or publish-subscriber middleware (more on this later in this chapter). Antoine de Saint-Exupery (French writer and aviator, 1900–1944) captures the essence of minimalism and what should be included in core functionality as follows: "Perfection is achieved, not when there is nothing more to add, but when there is nothing left to take away."

The minimalistic core pattern is directly applicable to mobile devices, services, and applications. For example, the Symbian operating system, an open mobile operating system with greater than 50% market share worldwide, is built using this design pattern. In Symbian, the user interface is separate from system and application services, and services use a request and callback mechanism for interacting with system resources. The Java location API (discussed in Chapter 3) is an example of a minimalistic core approach to location-based services. In particular, this API provides access to location information by supporting interactions with numerous positioning technology providers. Replacing or extending positioning technology does not require code changes to applications built using the API.

7.3 USER INTERFACE DESIGN

User Interface (UI) design is the single most important aspect of an application, service, or product that can make this application, service, or product successful. Take, for example, the popularity of Apple's consumer devices (e.g., iPod and iPhone). Apple has been extremely successful in designing devices that offer a user interface that is simple, intuitive, and elegant. When designing UIs for mobile applications one has to take into account several mobile device factors that are both unique and challenging. In particular, emphasis should be placed on the following:

- *Small Form Factor*: Mobile devices have limited display real estate. Trying to fit as many items as possible in such small displays will make the interface very busy and confusing. Therefore, being able to present only the necessary information and controls is of paramount importance. When relatively large amounts of data need to be shown, then being able to offer visual cues about the relative importance of the data becomes important. Furthermore, taking advantage of the multi-modality capabilities of some of these devices (smart phones, music players, etc.) is extremely beneficial in bypassing the small screen limitation. For instance, playing an audio clip instead of presenting a lengthy text box is more appealing in many cases. Another important limitation of many mobile devices is their constrained input capabilities. Most cellular phones have a numeric keypad, some soft keys, and a four-way D pad. These input options make UI design quite challenging. Mobile UI designers should pay special attention to this since a complicated input approach would result in application features not being used that often or not at all. Again, a multi-modal approach to the input problem has many advantages, especially when the device itself can support multiple modalities (e.g., speech recognition, gestures, and touch screen).
- *Internet Connectivity*: For applications that depend on accessing network components, such as Web sites, one should take into account slow transfer rates and weak connectivity due to coverage. As we discussed in the client-server architectural pattern, being able to handle these two cases in a graceful way is critical to both user experience and mobile application usability.
- *Device Capabilities*: Not all mobile devices are created equal. Some may have more processing power than others may. Others may offer the ability to install custom applications that run locally and may maintain state using local resources.

Yet others may be able to interact with other devices in their surroundings and take advantage of some of the capabilities of these devices (e.g., for display or communication purposes). Therefore, UI design should be flexible enough to accommodate as many devices as possible in order to be accessible to a large number of mobile users.

- *Access Plans*: Depending on the particular access plan associated with a mobile device, one needs to consider the available options for communicating with the device in the context of available communication channels and costs. For example, when an unlimited data plan is available, the data channel is the preferred communications channel for message exchanges and media streaming. When the data channel is not available, the SMS channel could be used for message exchanges, provided that SMS costs are low and controlled (e.g., unlimited SMS plan). When SMS is not an option, one may choose to use a data-over-voice protocol (e.g., such as the ones offered by Airbiquity and Navox) for transferring messages between the device and a back-end application utilizing the voice channel. In the case where the device supports multiple radios (e.g., cellular, Bluetooth, and Wi-Fi), the most cost effective communication channel should be selected.

In the remainder of this section we will discuss some common design patterns that are applicable to mobile UI design, include voice user interfaces.

7.3.1 Wizards

There are many UI examples where a user trying to achieve a particular task gets lost in the process because she is not familiar with the full complexity of the task and the various steps that need to be followed in order for the task to be completed. The wizard design pattern addresses such situations by guiding the user through the steps required for achieving a task, one step at a time. During each step, appropriate feedback may be provided to keep the user engaged and aware of what has been completed so far and what remains to be done. Furthermore, each task is linked to its predecessor so that the user can quickly go back to a previous step and revise some of the provided information.

The wizard pattern is especially suited for mobile applications where the small screen size makes form filling an awkward task when the number of input fields is rather large. Furthermore, this pattern is also applicable to tasks consisting of steps that may be executed under a different modality (e.g., text input, speech input, or gesture input). Figure 7-2 illustrates the generalized multi-modal wizard pattern for two modalities, speech and text.

A mobile application that can benefit from the multi-modal wizard pattern is personal navigation. Here, one can request turn-by-turn directions for a specific destination address by just speaking the address. If any part of the addresses is not accurately recognized by the speech recognition engine used by the navigation device, this part could be entered by just typing the correct information using the physical or virtual keyboard of the device.

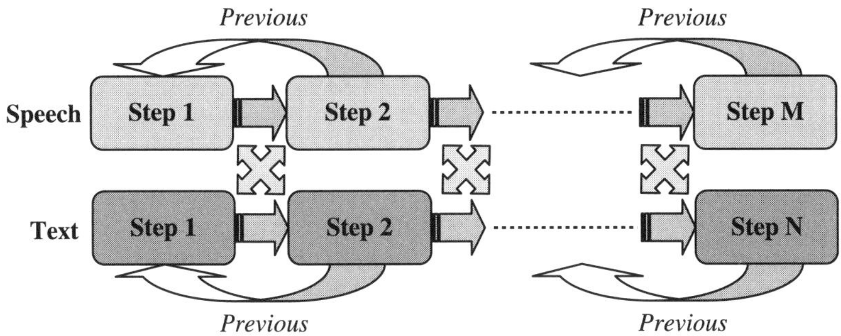

Figure 7-2: Multi-modal wizard design pattern.

7.3.2 Slideshows

Given the growing number of mobile phones that are equipped with an embedded camera, the ability to organize pictures taken by the phone's camera, and stored on the phone itself, into a slideshow that could be viewed on the phone or uploaded to a Web site and shared with others will become widely available. Mobile carriers have already started offering such applications to their mobile subscribers (e.g., Alltel Wireless MyShow).

However, when a user wants to navigate through a picture slideshow or some other series of media or information content on a mobile device, the screen real estate occupied by the displayed content and the available navigation controls are important UI factors that contribute to a pleasant user experience. The slideshow design pattern addresses such media or information content navigation by providing timer-based transition between items and making available controls for manual navigation (backward, forward, pause, stop, resume, and exit). Each "slide" is displayed for a specific duration and then the next "slide" is shown. The time delay between transitions can be fixed or configurable.

Depending on the specific features offered by the mobile device, different options may be available for controlling navigation. For touch-screen devices, a touch-based interface would do just fine. For devices supporting gestures, moving or shaking the phone in a particular way could be used for the same purpose. For devices with soft keys, these keys could be used for controlling navigation. Some of the numeric keypad keys could also be used in a way similar to the buttons found in DVD players. A speech-based approach could also be offered for controlling navigation. Here, a predefined list of voice commands is used for controlling slide transitions, direction, and speed. Table 7-1 shows an example command set for a speech-recognition based interface.

When a speech recognition interface is made available for slideshow navigation, one must ensure that the ability to disable voice commands and rely only on keypad entries (DTMF) or some other input mechanism is available. Without such ability, background noise would render the application useless since the speech recognition engine would not be able to correctly detect the entered command and would keep asking for input, resulting in a lot of frustration.

Table 7-1: Speech recognition based control of slideshows

Navigation Options	
Press	***Say***
* *	"Exit" | "Finish" | "Done"
1	"Rewind"
2	"Pause" / "Resume"
3	"Fast Forward"
4	"Previous"
6	"Next"
To disable Voice Commands: * #	

7.3.3 Pagination

The pagination pattern attempts to solve the problem of breaking up a large content into small chunks (pages) for offering a more user-friendly presentation. Once the large content is broken up into smaller pieces, navigation between them can utilize a variation of the slideshow pattern discussed in the previous section. Figure 7-3 illustrates several pagination examples being offered by popular Web and mobile services. Typically, the navigation component of pagination does not include a timer-based transition between pages. This is because several items may be displayed in one page at once. Again, depending on the capabilities of the mobile device, different options are available for implementing navigation between pages, such as touch, gesture, voice commands, soft keys, and numeric keypad.

Besides the breaking up of information into manageable chunks, the pagination patterns can offer efficient resource utilization on the mobile device itself, the

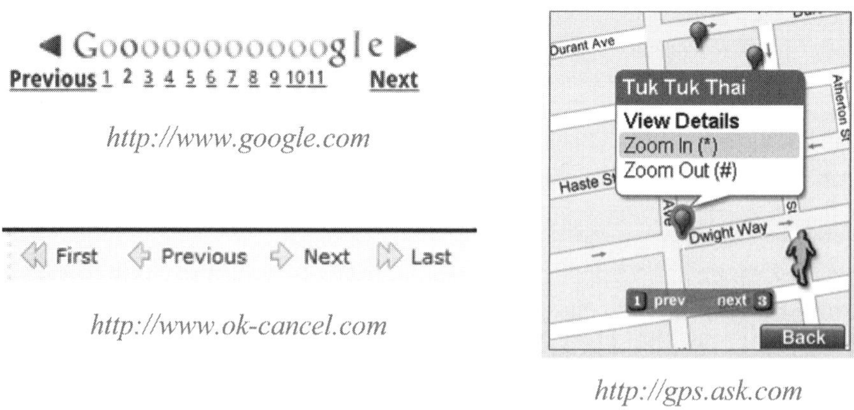

Figure 7-3: Pagination design pattern examples.

network, or the back-end service serving the information requested by the user. This is because only small parts of the entire information need to be accessed and processed for each page.

On the mobile device, pagination allows loading of the current page and, perhaps, several pages following or preceding the current page. By doing so, only a small amount of memory is required for holding this information. On the network side, the size of data transferred to the mobile device is relatively small. When the user accesses only a small percentage of the entire information, the savings in network traffic could be considerable. On the service side, only a small amount of information is retrieved and sent to the mobile device at a time, resulting in better resource utilization.

7.3.4 Cascading Menus

A menu is the primary way of navigating information that is hierarchical in nature, such as application menus and directory listings. When several levels of menus exist, organizing and accessing these menus in a convenient, efficient, and user-friendly way is an important aspect of UI design. This becomes even more crucial when the menu hierarchy needs to be navigated using a small factor device, such as a cellular phone or PDA.

The cascading menu pattern addresses this problem by adopting a variation of the MVC pattern we discussed in an earlier section. In particular, a tree menu model represents the menu hierarchy. The view component displays the menu options associated with the current state of the model. Finally, the controller component updates the state of the model based on user actions. Figure 7-4 illustrates the cascading menu pattern a simple city guide.

On a mobile device, menu selection can be performed in a number of ways. When a touch screen is available, touching the appropriate menu item will result in a transition

Figure 7-4: Example of the cascading menu design pattern.

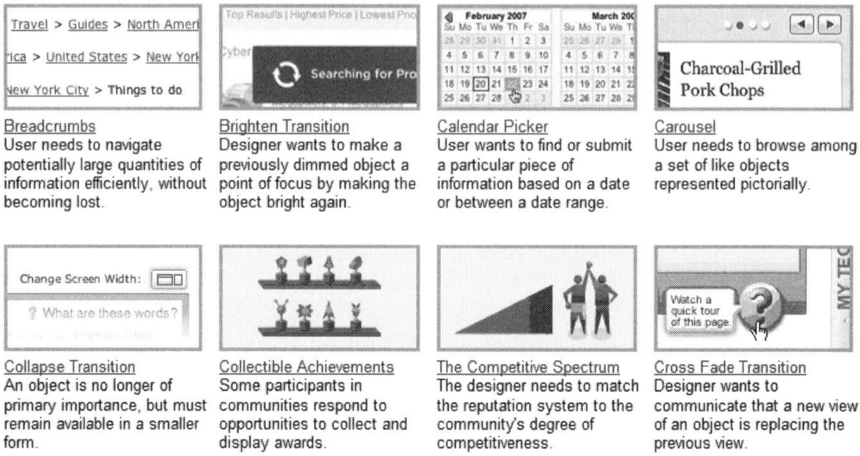

Figure 7-5: Visual and special effects "patterns" for Yahoo! Mobile widget developers. Reproduced with permission of Yahoo! Inc. © 2008 by Yahoo! Inc. YAHOO! and the YAHOO! logo are trademarks of Yahoo! Inc.

to the sub-menu associated with the selected menu. Similarly, soft keys, gestures, voice commands, the numeric keypad, or the QWERTY keypad could be used, if available, for selecting a particular menu item, going back to the parent menu, or accessing a leaf-level entry. Similar to the slideshow and pagination patterns, when a speech recognition interface is offered for controlling navigation, it is important to be able to turn speech recognition off and use a different input method (touch, numeric keypad, gesture) for menu navigation for privacy and handling noisy environments.

7.3.5 Mobile Widget Visual Patterns

While desktop widgets enabled by technologies such as Windows Vista, Yahoo! Gadget, Google Desktop, and Apple Dashboard are mature, there is some momentum to try to find the same success for the mobile device screen. A mobile widget, then, is a small program that runs on the mobile device screen and within a widget platform (which in turn runs on the mobile operating system). Mobile widgets typically offer simple and fun functionality, such as displaying customized weather or stock information, acting as a calculator, a clock, or almost anything else. A mobile widget can be developed to either run in isolation or communicate with other entities using the mobile device's data connection. Widgets are generally open source, or nonproprietary; some of today's platforms for mobile widgets include Yahoo! Widgets,[2] Plusmo,[3] WidX,[4] and iPhone.[5]

Figure 7-5 illustrates at a high level just a few of the visual and transitional "patterns" that are important to widget developers. These tricks and techniques can be

[2]http://widgets.yahoo.com
[3]http://plusmo.com
[4]http://www.joemoby.com
[5]http://iphonewidgetlist.com

considered patterns because they are used repeatedly in different development situations to solve challenges related to display and interfacing on the mobile device. Some of the patterns described in the figure as being available to Yahoo! Widget developers include:

- *A Carousel Effect* – The ability of the widget to lay out and move a set of icons in a virtual circular "carousel" effect (seen often on Apple iPods), and to allow the selection of one.
- *Transitions* – A transition is triggered upon user action. For example, after pressing a button that offers to hide some information on the widget screen, a transition pattern may brighten or darken the object until it disappears, or may "collapse" the object visually. Transitions help to provide a particular user experience and look and feel.
- *Calendar Picker* – A visual effect that renders a calendar pop-up, allows the browsing and selection of days, and returns the selection choice in a callback that the developer can use. Calendars are important in many contexts, particularly for mobile widgets related to notification systems (see Chapter 8).

Having a library of these patterns available to the developer for examination and direct inclusion into a mobile project is obviously of huge benefit to all parties. Having a community of widget developers contributing to a quorum of widgets that anyone can re-use, is even better, and this is the goal of many of the widget platforms today.

7.3.6 Voice Interface

Voice interfaces require short prompt-command-response interactions: the system plays a prompt, the user responds with a command, and the system then outputs the results. When the cycle is completed, the system initiates a new one. This model is quite different from the typical Web access model where a user selection is followed by the loading of a Web page, and the cycle starts over only when the user makes another selection. The Web access model lets the user pause indefinitely between navigation commands. By contrast, voice interfaces are not designed to work with idle periods or user-initiated actions; they require continuous interaction.

The prompt-command-response model pressures the user to respond within a particular timeout period. If the user exceeds that period, the system will time out and prompt the user again for input. The user might require additional time before responding, and the resulting prompts will most certainly annoy the user, perhaps causing her to terminate the call or exit the application. In other cases, the user might simply "get lost" somewhere in the call flow with no obvious way to return to a previously known state or the main menu. This can happen quite frequently to people who are accustomed to visual cues.

Therefore, a carefully designed and tuned voice user interface is of paramount importance to its usability. Since the voice interface may be used for authoring of content, browsing existing content, and searching, the call flow must support seamless navigation among authoring, browsing, and searching tasks, while providing extensive help support for all menus and shortcuts to frequently used parts of the service. The following interaction blocks are typically part of a voice user interface.

Greeting and Authentication The greeting block may be configurable to deliver a particular brand when the service is branded for an institutional environment. The authentication dialog is similar to those found in corporate voicemail systems – it forces users to establish their identity and lets them personalize the service.

Browsing Typically, the cascading menu pattern is used for browsing content using a voice user interface. Content is organized into categories and subcategories, and users navigate the category hierarchy by first selecting a top-level hierarchy by saying its name (e.g., "sports"), and then selecting a sub-category within this category (e.g., "soccer scores"). However, unlike the traditional cascading menu pattern that enforces a category hierarchy and allows navigation between parent and children categories only, a voice user interface can support a flat hierarchy. A user can access content associated with a category by just saying the name of the category, independent of where this category is located in the category hierarchy structure, provided that category names are unique.

This shortcut mechanism can be used to make personalized content categories accessible from the main menu (the user can speak the category name, such as "latest business news," from the main menu, instead of having to first select "business" and then "latest business news"). If the user does not immediately remember the name of a category, or simply wants to browse the list of available categories, she can just say something like "categories" to hear the names of the available top-level categories and select one. Once she selects a category, she can request the list of all subcategories, if any, by following the same approach, that is, issuing a voice command.

Authoring The authoring user interface is invoked whenever a user wishes to create a new media content or update existing content. Created content may consist of several parts, such as headline, main body, attachments, and keywords. Some of these parts may correspond to recorded user voice input and some of them may correspond to existing textual or multimedia content that is linked together with the audio parts (e.g., uploaded pictures). The authoring interface should guide the user through the steps of recording audio pieces, linking them with existing content, attaching keywords, and verifying the final product before incorporating it into one or more of the content categories.

The authoring interface may need to access the search interface in order to retrieve existing content for the purpose of updating it or reviewing it before making it available for others to access it (e.g., compliance process). Therefore, the design or the various voice menus associated with the authoring piece should support integration with the search menus in a transparent way and without confusing the user. Remember that when interacting with a service using a voice interface there are no visual information that could be used for reminding the user about the current context. Therefore, the authoring menus should incorporate sufficient feedback to aid the user in orienting herself during each step.

Search In most cases, organizing content into categories and subcategories is not sufficient for efficient access. By attaching keywords and tags to the available content, users can perform associative searches using the voice interface by just saying the keywords they are interested in, similar to typing keywords on a Web search engine. Of course, in order for this to work, content needs to be associated with metadata that consists of keywords and other tags. This can be done using the authoring interface, a metadata extraction application, or a Web site. The search application could be made more efficient by following a hierarchical structure in which each hierarchy level lets users narrow down their search by supplying additional keywords and constraints.

Delivery Once the system has selected a piece of content (either through the search application or by selecting a category), the content delivery menu is invoked for making the content available to the user. Unlike graphical user interfaces that can display multiple results on the same page, the voice user interface is serial. Therefore, a variation of the pagination pattern needs to be applied here, where each page contains only a single piece of content. The user can move forward and backward by issuing voice commands or pressing keypad buttons. In addition, the lack of random access requires the result space be constrained as much as possible; the user is unlikely to listen to 19 audio clips when the 20th is the one in which she is really interested in. To accelerate the process, the delivery interface must enable forward and backward movement within the same content (e.g., fast-forward and rewind functionality) or accelerate playback speed.

Help Without visual navigation marks, support for help dialogs is extremely important. Every dialog in a voice user interface should support the "help" command, identifying the current dialog and presenting the user with the available options. Ideally, the help menus should be customized to how experienced a user is with interacting with the voice interface. The amount of information included in the various help menus and number of help menus enabled at any point in time should be controlled by the user profile and voice or DTMF commands in real-time.

7.4 DISTRIBUTION AND RESOURCE MANAGEMENT

Developing distributed applications is a not easy due to the inherent complexities of distributed communications. In a mobile environment, distributed communications complexities have a higher impact on applications due to the nature of mobile communications: latency, speed, and weak connectivity. In this section, we will discuss some of the most common design patterns for distributed mobile applications.

7.4.1 Remote Proxy

The remote proxy pattern addresses the problem of interacting with a remote application that requires input, or produces output, that cannot be handled by the mobile

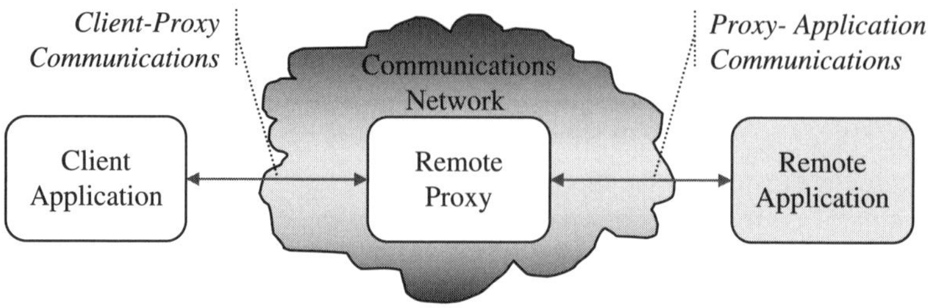

Figure 7-6: Remote proxy communication pattern.

device itself due to either lack of necessary resources or high costs associated with accessing such resources (e.g., battery demands, CPU utilization). The solution to this problem is to delegate interactions with the remote application to a network component that has the necessary resources for carrying out all interactions between the device and the remote application. Figure 7-6 illustrates this.

The following example makes the notion of the remote proxy pattern concrete. Let us assume that we want to access driving directions between two points using our cellular phone that is data capable. Under normal circumstances, we would just open the device's Web browser and access such directions from one of the available Internet sites that offer such services. Unfortunately, we happened to be at a particular geographic region where data services are not available. How could we still access directions in this situation?

A solution to the above problem is to use a remote proxy that offers a speech-based interface. Here, we would call a telephone number associated with the proxy, enter start and destination addresses using speech, and then listen to the audio directions offered by the remote proxy. The remote proxy uses speech-to-text software for collecting start and destination information, communicates with a Web site offering directions using HTTP (or some other Internet protocol), and then generates an audio version of the directions using Text-to-Speech software and let us control the playback of the content using speech or DTMF tones.

7.4.2 Data Transfer Object

The client component of a mobile application, which runs on a mobile device, may require access to several attributes of one or more objects by interacting with a remote server application that manages these objects. Several alternatives are available for enabling such accesses. On one hand, the client could issue a separate request for each object attribute. On the other hand, the client could fetch several or all object attributes at once and then access them locally. The former approach incurs significant network traffic since each attribute access requires a network roundtrip. The latter approach minimizes such network overhead by fetching all or a subset of the business object attributes in one network roundtrip.

Depending on access restrictions and business logic, different clients may be able to fetch different subsets of object attributes. In order to accommodate this, it is often the case that different objects are used for encapsulating the subset of the original object attributes a client is allowed to access. Instances of these objects are then retrieved by clients and accessed locally.

The above client-server interactions via the use of objects that encapsulate a subset of attributes belonging to one or more objects are represented by the Data Transfer Object (DTO) pattern, which is also known as Data Value Object (DVO). DTO has gained a lot of attention is the context of Java enterprise applications and is considered a core Java Enterprise Edition (Java EE) pattern.[6] Besides Java EE, the DTO pattern has been used in many other application domains [BIL].

DTO need to provide means for representing their state is a way that facilitates network transfer. Typically, there are two approaches to offering such functionality. One is to use a binary representation. The second is to use a textual representation. The former approach needs to take care of the byte order representation of multi-byte scalar types (e.g., integers or float), known as Endianness. The second approach needs to provide means to map object state into the appropriate encoding (e.g., XML, ASN.1, and JSON).

The DTO is extremely important for mobile applications that involve retrieval of information from multiple information sources, such as traffic and weather information for a particular city. As we have mentioned several times before, not all mobile devices are the same, and the network connection between a mobile device and a network server may exhibit different bandwidth and reliability characteristics. Therefore, it is important to be able to design the mobile application in a way that could handle DTO of different granularities. Larger objects can be used when the device is capable of storing these objects and the network connection offers enough bandwidth. On the other hand, more granular object should be used when network bandwidth is low or when the device does not have enough memory or storage space.

7.4.3 Remote Façade

In a client-server application where a client needs to fetch or update several properties of an entity, the client may have to exchange multiple messages with this entity, one for each property it needs to fetch or update. When the entity is at a remote physical location, each message is sent over the network and incurs overhead and potentially monetary costs (e.g., when SMS is used for message exchanges, each SMS may be charged). In order to reduce this overhead or costs, one may retrieve or update a number of properties with a single message exchange. This coarse-grained interface on top of fine-grained objects is referred to as Remote Façade. Figure 7-7 demonstrates this pattern in the context of a location-based application that retrieves GPS coordinates.

Since mobile applications are distributed in nature, reducing network interaction via the remote façade pattern has a high impact on the overall performance of the application.

[6]See http://java.sun.com/blueprints/corej2eepatterns/Patterns/TransferObject.html for details.

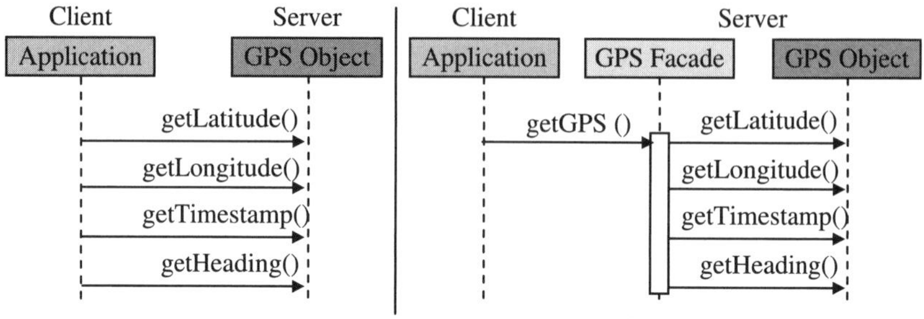

Figure 7-7: Remote facade pattern example.

7.4.4 Caching

Applications running on mobile devices are typically designed to work in two main modes: *connected* and *disconnected*. In the connected mode, mobile applications communicate with a remote server for carrying our application specific tasks, such as audio streaming or accessing data sources. In the disconnected mode, mobile applications rely on local resources only for their operation. When in connected mode, mobile applications do not need to maintain local copies of remote resources since they can always access them from the remote server.

However, since we have not yet reached a point where a reliable data connection is always available between the mobile device and the remote server and, in addition, always-on data connection may incur additional monetary costs, operating in a connected mode many not be a feasible approach at all times. In these cases, instead of not being able to use the mobile application at all, it may be preferable to cache remote resources to the mobile device and access them locally when in disconnected mode.

Depending on the nature of the cached data (frequently changed, mostly static, read-only, or read-write), different caching approaches may be followed. For example, an embedded database management system may be used for storing cached data locally in a reliable way. Furthermore, additional Web components may need to be installed locally, such as Web proxy servers. This is how Web applications can run when not connected under Google Gears.[7]

7.4.5 Pre-Fetching

In many cases, an application may be able to anticipate the next steps a user is going to take. For example, a personal navigation device knows the current location and heading of a vehicle and can predict where the vehicle is going to be in five minutes based on where the vehicle has been in the recent past and information about the road segments ahead. In such cases, being able to have the information required already available improves the user experience a great deal due to the perceived responsiveness of

[7] http://gears.google.com

the application. The pre-fetching pattern addresses this situation by allowing information to be pre-fetched and cached locally ahead of time.

In the case of mobile applications where latency and connectivity are not always constant, pre-fetching application data before the user accesses the data can have a significant impact in user experience. While determining the information a user is going to access next with accuracy is not always feasible for all mobile applications, there are several classes of mobile applications where deciding what to pre-fetch is straightforward. An example of such applications is e-mail. Here, pre-fetching of recent e-mails, especially new and unread e-mails, is beneficial in most cases.

Pre-fetching requires the use of both resources local to the mobile device and mobile network resources. The former may have an impact on the battery power and available memory of the mobile device. The latter may increase monetary costs associated with data accessed. Therefore, one needs to consider these factors before choosing to use the pre-fetching pattern [YIN].

7.4.6 Synchronization

Several classes of distributed mobile applications may have to operate under a weak connectivity model where the connection between a mobile device and a remote server may not be always on. There are several technical, business, and practical reasons why this may be the case. For example, in the area of automotive telematics, maintaining a connection between an on-board device and the back-end data center open all the time will drain the battery when the vehicle is turned off for an extended period.

Depending on the specific mobile application requirements, a mobile device may be required to collect data from its environment, keep track of collected data, and then transmit the collected data or some representation of the collected data (e.g., aggregate information) to a remote server when a connection is established, either on demand or at periodic intervals. This is the case with vehicle tracking related automotive telematics services. Here, location information is collected and then transmitted at periodic intervals.

In some other cases, a mobile application may maintain a local copy of data present at a remote server and make changes to the data. For example, one may store her contact list on a server machine (corporate, personal, or the Internet) and maintain a local copy on the mobile device. Then, changes may be made to the local copy (contact information updates, inserts, or deletes). These changes would need to be propagated to the server next time the device is connected to the server. The same is true when changes take place at the server and the local device copy needs to be synchronized with the state at the server.

The synchronization design pattern addresses the above problem of synchronizing state information between a mobile device and a remote server. In particular, the synchronization pattern advises the implementation of a state tracking mechanism at the device and the remote server. The state tracking mechanism should be able to keep track of updates made to local state, communicate these updates with its counterpart

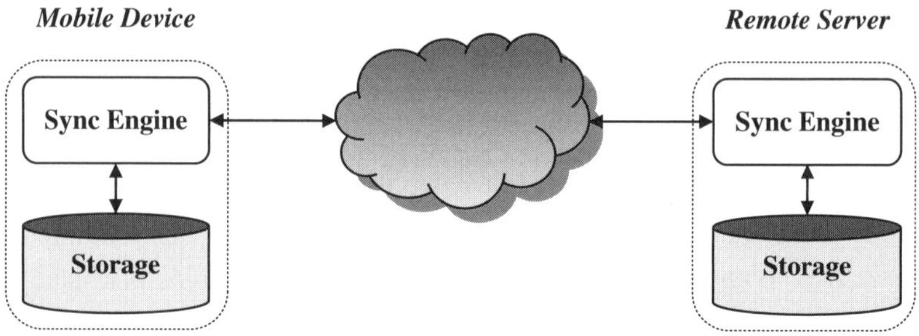

Figure 7-8: Synchronization design pattern.

at the remote end, and resolve any conflicts that may arise based on a well-defined conflict resolution mechanism.

Obviously, some kind of persistent storage must be available on the mobile device in order to record state updates. The persistent storage component can range from a file system to an embedded relational or object-oriented database. Figure 7-8 illustrates the synchronization pattern.

There have been several implementations of the synchronization pattern in the past. However, each of these implementations used its own data synchronization protocol that limited the number of both mobile devices and remote servers that could participate in state synchronization. In 2000, the Synchronization Markup Language (SyncML) initiative was formed to address the non-interoperable mobile synchronization solutions by aiming to create a single, common data synchronization protocol for mobile devices. The companies behind this initiative were Ericsson, IBM, Lotus, Motorola, Nokia, Palm Inc., Psion, and Starfish Software. Version 1.0 of the SyncML protocol was introduced at the end of 2000. Version 1.1 of the SyncML protocol was released in 2002.

An important property of the SyncML protocol is that several communications protocols are supported for synchronization message exchanges between a mobile device and a server device. One of these protocols, Object Exchange (OBEX), facilitates synchronization between devices using short-range communications technologies, such as Bluetooth and infrared. This allows mobile devices to synchronize calendar, contacts, and e-mail information with personal computers and laptops in a simple way. In fact, this is one of the ways in which mobile e-mail push vendors offer e-mail synchronization with popular e-mail software, such as Microsoft Outlook, including Web-based e-mail accounts.

Today, the Data Synchronization working groups of the Open Mobile Alliance[8] continues the work started by the original SyncML initiative under the OMA Data Synchronization (OMA DS) family of standards. Two of these standards, OMA Email Notification (OMA EMN) and OMA Data Synchronization Alerted Notification (OMA DS Push), are being used by existing mobile push e-mail providers for pushing e-mails or e-mail notifications to mobile phones. Both OMA EMN and

[8]http://www.openmobilealliance.org

OMA DS Push use SMS as their notification mechanism. Upon receiving such SMS messages the mobile phone establishes a connection with a mobile gateway for collecting the new e-mail.

7.5 EVENT NOTIFICATION

Most modern applications are distributed in nature and require support by services that can provide an infrastructure that facilitates:

- *Value-Added Services* – New services can be built based on events occurring at existing services without major modifications to these services.
- *Management Tools and Services* – The behavior of the system as a whole can be monitored by system users, administrators and developers.

The event notification design pattern (also known as event-driven or notification-based) is the foundation of event notification services that provide a key enabling technology for monitoring the behavior of large distributed systems, enabling reactive and proactive processing, and integrating independently developed applications. An event notification service accepts event descriptions from *suppliers* and delivers event notifications to *consumers*. Suppliers specify the kinds of events they supply while consumers specify the kinds of events that are of interested to them. Event notifications are delivered to only interested consumers avoiding wastage of shared resources.

Events are observable occurrences of interest in software/hardware components, data repositories, or the environment components interact with. Events occur at some specific point in time. Typically, events belong to two categories: primitive and composite. Primitive events are elementary occurrences of interest, such as state changes (e.g., current vehicle location, low gas, and airbag deployment), operation executions (e.g., creation of calendar appointment), occurrence of time, and application specific. Complex occurrence patterns consist of a combination of primitive events. Typical operators used for building composite events include: conjunction (e.g., vehicle travelling between exits 22 and 8 on interstate 287 south **and** accident at exit 14 with one lane closed), disjunction (e.g., an email is received **or** a phone call is being placed), and sequence (e.g., an SMS is received and **then** an acknowledgment is sent to the sender). Primitive events can be detected in several ways, such as sensor or application output. Composite event detection depends on primitive event detection and the semantics of the operators used.

The event notification design pattern is best suited for the design of distributed mobile applications due to the following properties:

- *Mobility Support* – An event notification service decouples event consumers from event suppliers. Consumers do not need to be aware of the sources or the locations of the event they subscribe to receive. Similarly, suppliers do not need to be concerned with who the consumers of their published events are, or where these consumers are located.

- *Disconnected Operations* – An event notification service can offer both push and pull modes of operation and quality of service guarantees that allow buffering and persistence of event notifications. These properties allow suppliers and consumers to operate in a weak connectivity environment.
- *Efficient Multicast Delivery* – An event notification service can exploit the intrinsic broadcast properties of wireless communications and offer efficient dissemination of event notifications to consumers.

In the remainder of this section we will provide a high-level description of the OMG Notification Service. Since the overall functionality of this service is similar to that offered by existing publish-subscribe platforms and event-driven architectures, such as Java Message Service (JMS) and Web Services Notification (WS-Notification), it is worthwhile learning more about its overall structure and functionality.

7.5.1 OMG Notification Service

The OMG Notification Service [NOT] offers a decoupled, publish-subscribe model of communication using the concept of *events* that originate at *suppliers* and are delivered to any number of *consumers*. Suppliers and consumers do not need to know about each other. Figure 7-9 illustrates the overall architecture of the Notification Service. The main types of supported objects are *event channels*, *administrative objects*, and *proxies*. Event channels receive events from suppliers and dispatch them to

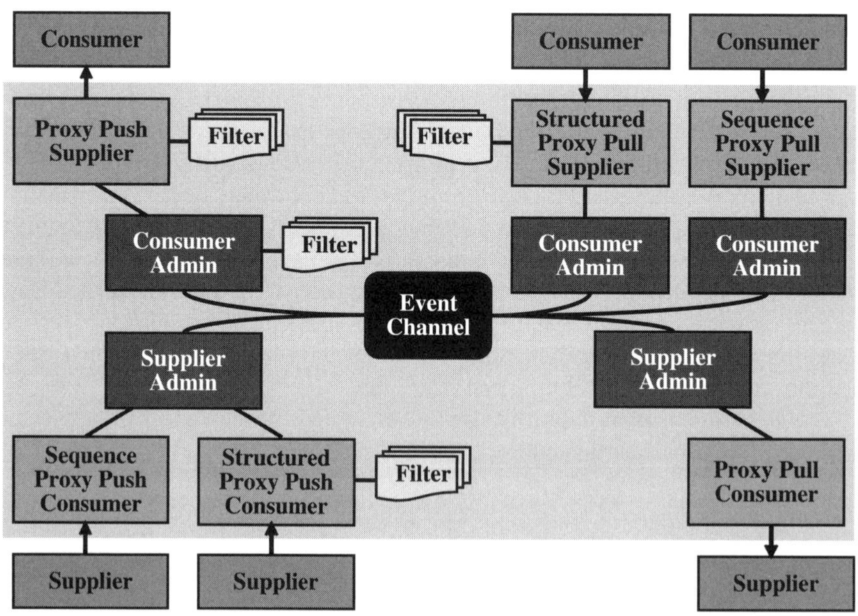

Figure 7-9: OMG Notification Service architecture.

consumers. Administrative objects are factories for proxy objects and provide "group-level" operations that apply to all of their proxies. Finally, proxy objects are the communication end-points for clients of the service, that is, event suppliers and consumers.

The Notification Service supports two styles of event communication: *push* and *pull*. A push event supplier sends events to a channel by invoking a push operation on the proxy object it is connected with, and the channel delivers events to event consumers by invoking a push operation on them. The channel requests events from pull suppliers by invoking a pull operation on them, and consumers receive events by invoking a pull operation on the proxy object they are connected with. Both operations are implemented using blocking or non-blocking two-way communications, and all combinations of push/pull suppliers and consumers can be used.

The Notification Service supports three kinds of events: *untyped*, *typed*, and *structured*, as well as sequences of structured events. Untyped events contain data structures defined by the clients of the service. Typed events correspond to specific interfaces used for event communication. Finally, structured events provide a well-defined data structure into which a wide variety of event types can be mapped, including XML representation of events.

Suppliers and consumers can associate *filters* with both proxies and administrative objects. Each filter contains a list of *constraints*. Each constraint is a Boolean expression, which is specified in a constraint language, over the attributes of an event. At least one of the filter constraints must be satisfied in order for an event to be sent to the channel or forwarded to a consumer. When many filters are associated with a proxy/administrative object, at least one filter must be satisfied in order for an event to be either sent to the channel or forwarded to a consumer. Filters attached to administrative objects are shared by all proxies created by this object. When filters are attached to both an administrative object and one of its proxies, these filters are combined using either AND/OR semantics. The case where no filter is specified is equivalent to a filter that is always satisfied.

An important feature of the Notification Service is the ability for clients of the service to define quality of service requirements that determine the reliability, validity, and delivery characteristics of event messages. The quality of service requirements can be specified at several granularity levels (event message, proxy, administrative object, and channel). Finally, the Notification Service defines interfaces that enable the creation and management of networks of channels. These networks can be configured through these interfaces so that they can support cohesive quality of service policies and implement an efficient event dissemination topology.

7.6 USE CASE: VOICE-BASED CONTENT MANAGEMENT

Today, we live in a world of constant change and explosive information growth. We increasingly rely on electronic creation, storage, and dissemination of information. We require access to relevant and time-sensitive information regardless of physical location, time of day or device used. As people get accustomed to the plethora of information and services that are available in the Internet, the transition to the telephone and other pervasive devices becomes natural. The mobile phone, in particular, is becoming an important platform for creating and accessing digital information. In the mobile phone world, people have access to information and can create and disseminate information when and where they want by using the most natural and trusted medium, the human voice.

In this section, we discuss the area of voice-based content creation and delivery and present a use case for capturing, classifying, and accessing content using a telephone. In particular, we discuss voice application development lifecycle, explain the information model, and present the overall system architecture that uses the latest technologies in speech recognition and voice markup languages (e.g., VoiceXML [VXML]) and utilizes several of the design patterns discussed in this chapter.

7.6.1 Voice Application Development Life-Cycle

The development process for applications the use speech recognition differs from that of most other telephony software in that the user interface is more complex than a simple touch-tone user interface designed for the same tasks. In addition, speech applications demand a formal deployment phase that occurs after functional testing is completed. The deployment phase introduces several rounds of "usability" analysis and tuning designed to improve the performance of the speech user interface. Application specification and development are the other two phases of voice application development lifecycle. A discussion of each stage follows.

Specification Phase The specification phase begins with requirement analysis for developing a preliminary user interface design that will identify the framework for building the speech recognition contexts (i.e., the grammars and vocabularies). Concurrent with the UI design, the functional components of the overall system design are identified (e.g., database and telephony systems integration), allowing two parallel development paths for the speech application development and systems integration. These two paths will converge in the next phase of the development life-cycle. The final output of the specification phase is a detailed project plan that identifies deliverables, dates, and responsibilities. Using a critical path method of project planning with minor and major milestones to track progress is recommended in this phase. Figure 7-10 illustrates the steps of the specification phase.

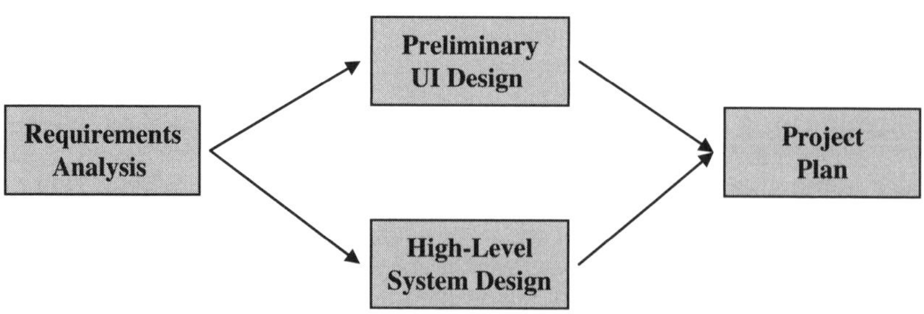

Figure 7-10: Voice application specification phase.

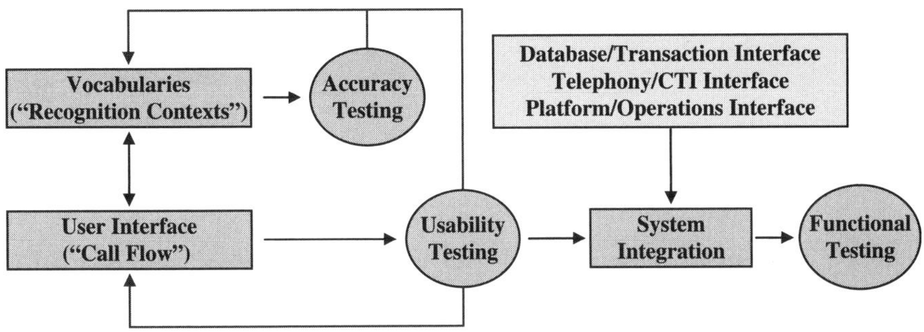

Figure 7-11: Voice application development phase.

Development Phase The development phase represents the creation of the speech application completed in a series of rapid prototyping loops and the interaction with host databases and telephony middleware. These two paths converge at the systems integration block, representing the first point where the application is ready for functional "end-to-end" testing. The output of the development phase is a completed application, integrated with the database transaction engine and the telephony network that has passed a rigorous series of functional tests. The functional tests include both user interface and application validation. Figure 7-11 illustrates the steps of the development phase.

Deployment Phase This phase addresses performance testing and user interface and recognition accuracy tuning. Here, the fully functional voice application is available to end users. Based upon user experience and collection of spoken utterances, the voice user interface and recognition models are tuned in order to balance the highest possible transaction completion rate and the shortest possible call duration. Figure 7-12 illustrates these steps.

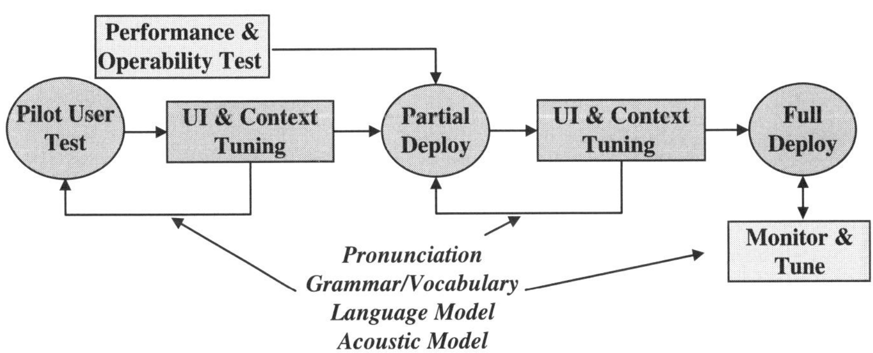

Figure 7-12: Voice application deployment phase.

7.6.2 Why Voice Authoring?

The fast proliferation of mobile devices is creating a new opportunity for the user-generated content management industry [ANE]. However, content in textual form is not easy to create and consume using small mobile devices. The small screen size makes it especially cumbersome to create and access lengthy content (many navigation/scrolling commands, small fonts, and inconvenient keyboards). A voice interface does not have these limitations. Firstly, it is available through today's most pervasive device – the cellular telephone. Secondly, recent advances in speech recognition technology provide more efficient voice capture interfaces and indexing mechanism (in contrast to touchtone interfaces that are more cumbersome than small screens and keyboards). Thirdly, humans can speak faster than they type. Finally, voice authoring can be preformed anywhere, without requiring much attention on hitting the correct keys on a keyboard or finding a comfortable position.

In an environment where the time between content creation and delivery is critical, voice authoring has significant advantages. It reduces the time necessary to capture audio content without increasing the time required to index it properly and empowers users with content creation abilities when mobile. Some arguments in support of voice authoring were presented in [CHA], where users indicated their preference in using their voice to express comments during an experiment evaluating textual and voice annotations. Several other studies have also found that voice is more expressive than text. A speaker can communicate much more information through speed, pitch or tone – something that is not possible with text.

There has been a good deal of research around the general capture and access theme. An excellent review can be found in [ABO]. Most of that body of work is related to capturing raw audio and video from classrooms, meetings, etc., and applying techniques either during or after the capture to organize it appropriately. Hindus et al. [HIND] addressed more closely the issue of audio content by capturing telephone conversations and spoken collaborations in work environments and deriving some simple structure during or after capture. An important aspect of voice authoring is the ability to attach meta-data to the captured audio. Fields such as author, title, creation date, keywords, etc., are particularly useful in creating a comprehensive searchable index of the captured content. Schmandt et al. in [SCHM] and Resnick in [RES] have demonstrated the value of generating semi-structured audio (i.e., raw audio with the associated meta-data) using form-entry metaphors.

7.6.3 Information Model

The fundamental unit of content is a *story*. A story consists of media components, called *story items*, and pertinent metadata. Story items are multimedia files (e.g., audio and video files, textual documents, images) that may exist in several formats to support different access devices, such as telephone and Web (e.g., as G.711, Windows Media, Real Audio, and MP3). Story metadata is textual and consists of fixed attributes (e.g., title, timestamp, length, state) and name-value pair keywords. Content administrators define the categories to be used as keyword names

Table 7-2: Information model – example story content

ID	01JX2008082508SL		
Title	My recent trip to Dominical Republic		
Author	John Smith		
Date	08/25/2008 01:08 PM		
State	Online		
Story Items	***Name***	***Length/Size***	***Media Format***
	Body	2:11	G711
	Body	2:11	WMA
	Santo Domingo	1.01 MB	JPG
	Isla Saona	1.22 MB	JPG
Keywords	***Country***	Dominican Republic	
	City	Santo Domingo, Isla Saona, Saona Island	
	Currency	Peso	

(e.g., region, state, and city). Keyword values are selected from enumerated or unrestricted sets. Speech recognition can be used for the former because they translate into finite vocabularies. The latter are more suitable for Web access. Table 7-2 shows the story components according to the above information model.

In many cases, audio content needs to be combined with external data to satisfy the requirements of a specific service. For example, a story about a traffic accident may be linked to both an external traffic service provider for retrieving up-to-date information about the accident and a turn-by-turn traffic directions provider for accessing alternate routing information. The ability to link authored content with internal or external media, real-time feeds, and services is very powerful. The "wizard" design pattern plays a central role in helping a user carry out such content linking using a phone, Web browser, or a combination of both at the same time.

The "cascading menus" pattern is used for content organization. In particular, stories are grouped into *channels*, which represent thematic grouping structures. For example, an "Automotive Industry" channel will contain news and research on issues related to the automotive industry. A story is always created in the context of a specific channel. After its creation, however, a story can be associated (via a Web interface) with more channels, subject to access restrictions. Channels are then organized in a structure of categories and subcategories, very similar to how books are organized in a library. For example, the top-level categories could be "News," "Sports," etc. Under "News," several subcategories such as "Breaking News," "Financial," "Politics," and "Local" can exist. The category structure is determined by the content administrator and facilitates the user interface for navigating and locating stories for both PC (Web) and mobile phone access.

Channels can be of different types. The most common type is the *story channel*, which supports the creation and organization of media content in the form of stories. A *stream channel* allows live connections with Internet-based multimedia streams and feeds. Channels can also be linked to other applications such as various kinds of productivity applications (voice access to email, calendar, etc.). Depending on the access mode (phone, Web, etc.) a channel and the associated application are rendered by a mode-specific UI module (an interactive voice dialog for phone access, a Web page or form for Web access, etc.). These mode-specific UI modules correspond to the view component of the "model-view-controller" pattern and utilize the "pagination" and "slideshows" patterns for simplifying content accesses.

For authentication purposes, users are assigned unique numerical ids and passwords (general alphanumeric ids are not handled very reliably by speech recognition systems and are cumbersome to manage from devices with numeric-only keypads). Because the number of available categories and channels may be large, users can set up personal profiles that contain up to nine ("9") top-level categories with up to nine channels in each category. The choice of the number "9" may seem awkward initially, but in fact, it allows for a very efficient design of the voice interface (a bounded number of choices, with numeric fallback when touch-tone entries are desirable). Web access to the same profile is also more intuitive since the user has immediate access to the categories and channels that are most relevant to her. Profile management functions are offered through a Web site that allows the user to add and remove channels from their profile.

For security purposes, content and applications are subject to access control. The access control model is based on the concept of closed user groups, similar to access control lists. A closed user group consists of a list of users and a list of channels. Every channel has associated with it a list of access privileges for the users in that group. One or more of the above access privileges can be assigned to the users of a closed user group. When a user accesses a channel, her access privileges are computed as the union of privileges of all closed user groups that are associated with this channel and contain this user.

7.6.4 System Architecture

The information about users, closed user groups, stories, story items, keywords, channels and categories is stored in a relational database. Physical media files are stored in a suitably organized file system. A multi-modal middleware layer implements all required data management functions associated with these repositories. This middleware layer has three major components: a core API that provides access to database services, a media services API that performs physical media management functions (transcoding between media formats, managing media streams and broadcasting, converting text to speech, etc.), and the notification API that provides outbound user notifications. The middleware layer utilizes the "minimalistic core" and "client-server" design patterns for encapsulating specific functionality into components and supporting inter-component communications (e.g., Media Access coordinates accesses to content and content meta-data). In addition, the "caching" design patterns is used for minimizing disk accesses by caching frequently accessed content and meta-data in memory. Figure 7-13 illustrates the system architecture.

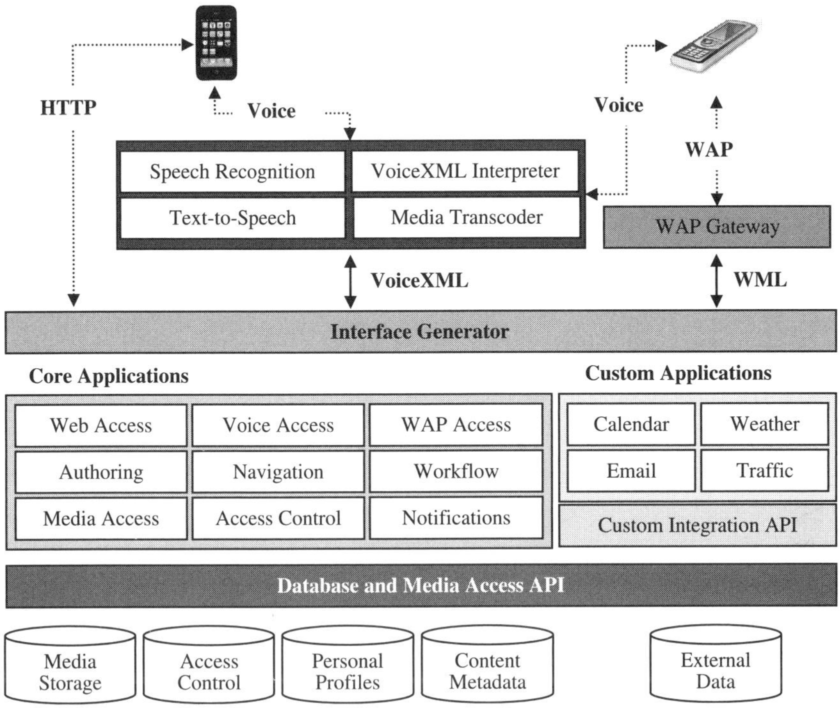

Figure 7-13: High-level architecture of voice authoring platform.

Telephones are primarily alerting devices and, thus, they are uniquely suited to support asynchronous delivery of information. As part of a personalization process, users can create filters for alerting them when new content is either created in a particular channel or authored by specific user, or when it contains specific keywords. Users can create multiple filters and manage existing filters using a Web interface. As part of this interface, users can create new filters, delete existing filters, suspend and resume filter evaluation, change filter-matching criteria, or specify the time window during which alerting is permitted for minimizing the intrusiveness of the service.

The notification service receives story creation events, evaluates registered user filters that are currently active, and creates a delivery list once a match is found. The delivery list indicates the user accounts that will receive an alert, subject to access restrictions. Depending on the user alerting preferences, the alert can be sent as a text message over e-mail or SMS, or through a telephone call (in which case the actual audio story items can be delivered). In addition to user alerts, the notification service can deliver text or phone alerts proactively to predefined lists of recipients. This feature is typically used by the author of a story to automatically alert a number of recipients on demand (e.g., members of a social network).

In this case, the recipients do not have to be users of the system, and the only information required for delivering such alerts is their e-mail addresses or telephone numbers.

7.7 SUMMARY

In this chapter, we covered a wide range of design patterns associated with mobile application development. In addition, we presented a voice-based content management architecture in order to draw attention to some of the technical details surrounding multi-modal services, and in particular services that use speech recognition as one of the primary user interfaces. Having digested all this information, the following questions should come to mind with respect to what the future may hold for mobile devices and services.

QUESTIONS

1. If speech recognition were offered as a standard component of mobile devices, would we need multi-modal interface support in back-end services?
2. If mobile devices use full-fledged operating systems with full support for Web browsers, HTML, and XML, would we still need to worry about content rendering on mobile devices at the back-end?
3. If mobile devices are capable of utilizing resources available in their surroundings, such as LCD televisions, or are able to project on different surfaces, would we still have to worry about small form factors?
4. If mobile devices become the computing platform of the future, would anybody develop applications for non-mobile devices?

8

ADVANCED SERVICES OF TODAY AND TOMORROW

"As we look ahead into the next century, leaders will be those who empower others."
—Bill Gates

The "mobile mindset" – similar to the "Web 2.0 mindset" – values experience and is willing to pay for it. Increasing user expectations as well increasing competition along the value chain have had the effect of commoditizing mobile services. In the commodities market, a business sells products that are undifferentiated, namely, have no unique features that it can charge extra for. In their book, *The Experience Economy* [PIN], Pine and Gilmore described an economy that replaces our "service economy" in which the "experience" or "memory" is what customers would be attracted to, and willing to spend money on. For example, they spend on the experience of drinking a cup of Starbucks coffee, not just the cup of coffee itself, which perhaps can be bought elsewhere for less. This trend in consumer behavior is one of the major forces shaping the future of mobile services. It is still too early in the game to see where it is all going to settle as opposing forces are in action across the value chains (see Chapter 2). We see the rising value of experiential mobile service already embodied in things like:

- "Immersive" (or at least 3D) virtual worlds ported and playable on the mobile device in such a way as to allow the user to experience and integrate both the virtual and real (e.g., voice sessions) into a single smooth experience.

The Fabric of Mobile Services: Software Paradigms and Business Demands, by Shoshana Loeb, Benjamin Falchuk, and Euthimios Panagos

- Mobile context-aware services that adapt to particular environments in order to bring targeted information to the user (such as museum navigation).
- Personal, ubiquitous, and ambient services that are all at once: tailored to the given user, available at all times, and communicative via "gentle" ambient cues (but at the right place and time).
- Video games that extend into the real world by emitting emails and SMS to players.

All bets are that innovative mobile services that deliver reliable, timely and relevant information to mobile users will tend to be of high-value and have good chances of user adoption. Today, a large number of mobile applications allow users to send and receive information within their social networks, however, trends seem to indicate that the cross-indexing and correlation of complex information in order to deliver higher impact "notifications" to the mobile user will also be of great importance. Apple and Google are both keenly aware of supporting these sorts of long running applications on mobile devices and both have designed for, or will allow, "relay architectures" (i.e., Apple *Notification Architecture* and Google's Talk). Higher impact notifications are "smarter" than prescripted ones, and are based on contextual integration at a deep level (rich semantics, artificial intelligence, machine learning, and expert systems are the tools of the trade here).

Today's telematics use cases, intra-vehicle, inter-vehicle, and vehicle-to-server computing, continue to evolve at a rapid pace and enable information services for both users and other networked computing systems. Intelligent fusion of telematics information with other aspects will allow the computing system to not only find and recommend gas stations before the vehicle is in crisis, but also choose the station so that the "diversion" costs from the user's current route are minimized. Similarly, using the mobile infrastructure to assist users in managing their wellness and health information is very topical. When the pragmatics of secure gathering, storage, and sharing of medical information have been addressed, many compelling mobile services can be developed. We are already seeing consumer demand for "personal trainer" applications that run on mobile devices, but much more is possible.

In the remainder of this chapter we expand what was discussed so far in the book and re-examine some of the topics in light of current and future trends in mobile services. In addition to the topics mentioned above, we look at what it takes to "get personal" with mobile users, emerging application areas that are seeing a good deal of activity and innovation, and interesting implementations of technology that hint at the fabric of tomorrow's mobile services.

8.1 EMERGING MOBILE APPLICATION AREAS

In the following passages we review several application areas that we believe are being transformed by mobility. This includes personal reminder systems, automotive telematics, mobile healthcare applications that enhance quality of life and care, mobile marketing, which is all about getting user's mind share and making the right offer at the right time and the right place, and virtual worlds with a real world physical mobility. While other mobile application areas exist (e.g., photography-related, urban games such as geocaching, peer to peer device interactions, and many more) we have chosen an interesting and illustrative subset.

8.1.1 Reminder Systems

Reminder systems are in a renaissance. Years ago, the "push technology" fad linked content providers directly to the desktop of the user's Internet-connected personal computer. When information was available, a communications session was established, or an existing one was reused, between the server and the computer and information was pushed to a client program on the computer. Upon arrival, it was displayed in various ways, but primarily on the desktop where it was presumably ubiquitously accessible. PointCast Inc. was one of the potential stars of this realm. Its service, like its imitators, was eventually doomed due to: a) its failure to capitalize on investment deals, b) use of proprietary software on the computer (instead of a Web browser), and c) technical glitches – such as slow speeds – that irked customers. Today's "almost always-on" mobile handsets may be the new screens of the PointCast generation. A main difference is that there are multiple ways to reach these mobile devices. Instead of a single proprietary connection, a server can hit the mobile device via:

- The phone/telephony infrastructure/cellular network;
- SMS channel;
- Data channel/Packet networks;
- Voice-mail.

In addition to the above enablers, it is easier than ever to create a Web presence – both the tools and the expertise of developers have improved. An online offering can now be built using primarily open-source software (such as the LAMP approach[1]), thereby dramatically lowering the barriers to entry. As a result of these and other new enablers, a series of single purpose and fun services have emerged that allow users with cellular phones to create, store, manage, and get notified of information on various dimensions. RememberTheMilk, PingMe, MonkeyOnYourBack, reQall, and OhDontForget are just a few of the many startups trying – at the time of writing – to make a run in the "reminders" space. Google, Microsoft, and Yahoo, of course, offer online calendar systems whose events can be preceded by reminders through SMS, email, and so on. The startups listed above are different from the now traditional online calendar because they have certain playfulness and a fun-factor and encourage messaging amongst one's social network.

[1]LAMP is an acronym describing a software stack, composed of open source projects: L = Linux operating system A = Apache Web server, M = MySQL database, P = PHP/Python/Perl scripting languages.

Another aspect of both mobile and desktop notification services is the ability to parse free text. For example, Google's online calendar service (GCal) presents a single textfield for input, rather than a set of form elements such as check boxes, radio boxes, pull down information, and buttons. The result of this is a streamlined immediate-mode way to input information, just click on a calendar day and start typing (e.g., "7 pm Dinner at Pancho's"). This turns out to be an effective approach not only for desktop calendars users but for mobile users as well. It minimizes the number of clicks and saves keyboard usage by allowing "coded messages." What is the code? GCal users need only enter a short phrase that answers the "what," "who," "where," and "when" of the appointment (only the "what" and "when" are actually required). In order to parse a user's newly entered appointment, GCal essentially assumes that the event will be described in the order above – it also pulls out the "where" by searching for the "in" or "at" token in the phrase and then taking the next token(s) as the "where." Similarly, the presence of "at" or "on" greatly help the parser pull out the time and/or day of the appointment, and the token "with" is understood to precede a "participant." For example, "pizza with John at Pizza Hut Monday 7 pm" conforms to the what/who/where/when ordering and the "with" and "at" tokens allow GCal Quick Add to parse and store the event.

Another interesting company was Sandy,[2] primarily a desktop Web application that sent reminders in a somewhat automated and intelligent fashion. Though the service is now closed, its intellectual property is now owned by Twitter. It interworked with mobile users, allowing both input (of appointments) and outbound notifications via SMS. Sandy encouraged registered users to send all appointment and task information toward its server, which could be done as simply as CC'ing the sandy email in all of one's own emails. When Sandy received such an email, it resolved the sender to a registered user and then proceeded to parse the contents of the message. The results of parsing were either to store some information for later retrieval or set a reminder for some later time. Like some of the other reminder sites that desire additional page-hits from their subscribers using their mobile devices, Sandy accepted inbound information in a shorthand notation. The shorthand made the service both more viable and attractive to mobile-only users. Some examples of the shorthand include:

- "Remind me to move the car in 15 minutes" – sets a reminder (via a pre-set method, such as SMS or e-mail) for 15 minutes from the current time.
- "Remind me I have a dentist appointment at noon @reminder in 30 minutes" – sets a reminder relative to the start of an appointment.
- "Remember Dad's number is 503-555-1212" – adds a new contact.
- "Remember groceries @weekly" – sets a weekly reminder to do groceries; there may be an additional "list" of items associated the task labeled "groceries."

Thus, for the mobile user, Sandy shorthand is very convenient and allows the mobile user a quick way to create a reminder. Other interesting built-in notations include the

[2]See http://iwantsandy.com

ability to use the "@" sign to represent recurring reminders (e.g., "@Saturday" means "every Saturday"), and to use relative times and dates in reminder settings (e.g., "in 20 minutes," "in 3 days"). In general, mobile shorthand and "auto complete" mechanisms are a good idea – many studies have shown that typing error rates within SMS messages are alarmingly high. Other studies show that touch-based keypads such as those in the iPhone, also may induce errors for those used to tactile keypads.

8.1.2 Automotive Telematics

In recent years, automotive Telematics Service Providers (TSPs) have been focusing on customer satisfaction and loyalty by offering a wide-range of information services that go beyond safety and security. These TSP services aim to enhance both the quality of driving experience and the overall user experience by offering seamless integration with existing customer relationship management (CRM) systems, providing location-based traffic, navigation and information services, offering emergency response and roadside assistance, and catering to personal needs and preferences by incorporating speech recognition and supporting mobile devices such as cellular phones and music players.

Despite these efforts, however, the majority of available telematics services are mostly vehicle-focused. Vehicles are equipped with on-board units (OBUs) that are responsible for human-to-machine interactions (HMI), access to vehicle-related information such as diagnostic trouble codes (DTCs), and connectivity to back-end services and call centers. While most of these services may offer Web-based access to some of their functionality outside the context of a vehicle, such as pre-trip planning, they fall short in fully embracing mobile devices carried by drivers and passengers. Moreover, virtually all these services incorporate their own, often proprietary, communications mechanisms between OBUs and network services and ignore the pervasiveness of mobile devices and the rapid evolution of mobile and personalized services.

Only recently automobile manufactures and TSPs have started offering integrated support for consumer electronics devices such as music players (e.g., Apple's iPod, Microsoft's Zune) and cellular phones. In some cases, such as the SYNC[3] offering from Ford Motor Company and Microsoft, the cellular phone is the primary communications mechanism between the in-vehicle infrastructure and the back-end services such as remote diagnostics and emergency communications in case of airbag deployments. We expect that future TSPs will fully embrace consumer electronics devices and allow two-way interactions between such devices, vehicle infrastructure, and network services. To make this happen will require overcoming the great discrepancy between the life cycle of automobiles and consumer electronic devices. Currently the median life of a car is 9.2 years and the design cycle is about 3.5 years, while consumer electronics have about 2.0 years of median life and 0.5 years of development cycle.

Another trend that seems to be gaining momentum is the integration of broadcast data (satellite or radio) with built-in navigation devices in order to offer one-way

[3]http://www.syncmyride.com

real-time data access in a cost efficient way. For instance, Ford is integrating the SYNC service with the SIRIUS Travel Link data service. The SIRIUS Travel Link service provides real-time traffic, weather, fuel prices, sports scores and movie listings. The broadcasted information is tagged with appropriate location indicators in order to enable intelligent filtering based on the current vehicle location (GPS coordinates). While in the case of Ford and SIRIUS custom navigation devices are required for both receiving the broadcasted data and filtering the data based on vehicle GPS coordinates, we expect that APIs will be made available for integration with consumer devices other than the built-in navigation unit.

Competing with auto manufacturers and TSPs are Personal Navigation Device (PND) manufacturers. Today, PNDs offer much more than just driving directions from point A to point B. For instance, they offer Bluetooth connectivity with cellular phones for hands-free calling and accessing back-end services, built-on MP3 players, speech recognition and text-to-speech, a large database of Points of Interest, and access to online content and information in real-time (such as gas prices, movie listings, etc.). In addition, PNDs have started embracing user context and linking with popular social network sites. A recent partnership between Garmin and uLocate is evidence that PNDs are indeed expanding their role from user-agnostic stand-alone devices to user-centered connected devices. We expect this trend to continue and expand into the handset-based navigation market.

Another development that is gaining traction is that of predictive traffic modeling that is being developed by existing traffic information service providers, including Journey Dynamics, Inrix, and TrafficCast. Predictive traffic modeling takes into account historical traffic patterns, real-time speed information, scheduled events, weather forecasts, vehicle type, and driving behavior, among other parameters, in order to provide more accurate information about travel conditions and expected arrival time. The impact of this technology could be quite extensive as it can incorporate timing constraints associated with user tasks (e.g., calendar appointments) in order to provide personalized routing.

Another exciting development in telematics is that of integrating in-vehicle infrastructure and communications with infrastructure and communications available in the environment surrounding the vehicle. For example, Streetline[4] is deploying wireless sensor technology for creating "smart" parking spaces. Parking spaces are equipped with sensors that detect whether a vehicle is currently occupying the space or not. This information is then sent to a back-end server in real-time using wireless communications. Telematics services can then access this information in real-time and inform drivers about parking availability near the current vehicle location or some other location.

Telematics services are becoming increasingly compelling. A general trend of integration with personal and vehicle-external information is fairly clear, as is the promise of next-generation navigation and in-vehicle information systems.

[4]A company focusing on addressing city management and operations problems via sensing solutions. More details are available online at: http://www.streetlinenetworks.com.

8.1.3 Mobile Healthcare Trends

The fast pace and stresses of modern life, the ubiquity of communications tools, and the poor dissemination and education of healthy lifestyles has caused a glut of issues for this generation, including increases in incidents of heart disease, diabetes and obesity – even in children. However, the ubiquity of the mobile phone and communications networks has also given rise to telemedicine. The American Telemedicine Association (ATA) defines telemedicine as, "the use of medical information exchanged from one site to another via electronic communications to improve patients' health status. Closely associated with telemedicine is the term *telehealth*, which is often used to encompass a broader definition of remote healthcare that does not always involve clinical services. Videoconferencing, transmission of still images, remote monitoring of vital signs, continuing medical education and nursing call centers are all considered part of telemedicine and telehealth."[5] It is widely agreed that the telehealth home market will continue to grow at a rapid rate, with some expecting the global telehealth market to exceed $8 billion by 2012.

These figures give rise to a sort of perfect storm in support of telemedicine: customers benefit, carriers benefit, and device-makers benefit. Medical practitioners presumably benefit as well by way of timely information delivered to them through monitoring centers, though there has been some debate in their circles over the pragmatics of remote monitoring, liability, insurance, and billing issues. There are also legal jurisdictional issues, political concerns about patient data privacy, and debate over who owns patient data. The introduction of modern mobile devices – including smartphones – into the telemedicine use cases is already underway in both research and practice.

On the medical records front, Google and Microsoft are beginning what might turn out to be influential plays into personal medical records storage and management. The *Google Health* and the Microsoft *Healthvault* (supported by *RelayHealth*) plans are important steps towards centralized storage and access to medical records on a personal level. Neither service has a specific play for mobile devices except that the records systems are accessible to most smartphones via on-device Web browsers.

On the network and operator front, there are a multitude of initiatives that indeed seem to prove that operators see possible profitable results from offering health-related services on their network. CardioNet.com is a company supplying wireless heart-monitoring equipment and services. Heart patients wear customized sensors which interact wirelessly with a portable device. When the device detects an abnormal condition it uses a cellular modem to transmit data to a central server where further policies and experts analyze the data. CardioNet.com and Qualcomm have been in a partnership for quite awhile in which Qualcomm provides the CDMA communications infrastructure over which ECG data from patient sensors gets transmitted to the CardioNet.com monitoring center. Qualcomm itself is planning to roll out an MVNO for health and fitness in 2009. It will feature various monitoring, advice, logging, and reporting capabilities over the wireless network and will interwork with the mobile handset, making time and place irrelevant to access important personal health information.

[5]http://www.atmeda.org/news/definition.html

Hitachi has developed and plans to market a wristwatch form of health monitor able to gather and synchronize health signs via the Internet (no cellular yet). Patient acceleration, pulse, and temperature can be sensed. IBM has also performed similar research and prototyping along the same lines. Telzuit adapts the Palm Treo smartphone and transmits ECG sensor data via Bluetooth®. The phone uses Cingular's network to send critical cardiac information to Telzuit's data center where it can be forward to practitioners and family members. Motorola's MOTOHEALTH is a general-purpose health reporting infrastructure, an "informatics system that enhances information exchange between end-users and their third-party care providers. It has the potential to reduce healthcare costs while increasing compliance to the care regimen for chronic disease monitoring."[6] Vodafone has ventured into patient compliance territory through an initiative in which short messages are sent to patients' mobile handsets in order to remind them of regimens, intake and prevention. Since reminder based systems that touch only the SMS plane require little infrastructure besides servers and SMS gateways, there are now several companies that are capable of achieving similar things, and don't need to operate the radio access network to do it. Compliance for Life™ and Leap of Faith® Technologies are two such companies.

The American Health Information Community is a group of stakeholders who provide recommendations to the Department of Health and Human Services (HHS) on making health records digital and interoperable, and provide assurance that the privacy and security of those records are protected. The AHIC creates IT use cases, such as:

- Remote Monitoring;
- Patient – Provider Secure Messaging;
- Personalized Healthcare.

and many more. These IT use cases are more and more sensitive and accommodating to the fact that people are mobile, in remote settings, and have both wired and wireless communications networks available to them. Much thought can and will be put into how new wireless technologies, devices, and standards will benefit HHS.

In summary, many companies on the health, wellness, and telemedicine value-chains are positioning for profitability. A large market for wireless biometric sensors that wirelessly convey their data to mobile handhelds has emerged; these devices exploit the handhelds' access to the cellular network in order to transmit sensed data to a server operated by a healthcare provider or data intermediary. Many of the major wireless infrastructure operators are investigating or already positioning services related to telemedicine. Even international organizations are seeing the potential of wireless telemedicine. The Phones for Health initiative, for example, is "a cutting-edge $10 million public-private partnership, which brings together mobile phone operators, handset manufacturers and technology companies – working in close collaboration with Ministries of Health, global health organizations, and other partners – to use the widespread and increasing mobile phone coverage in the

[6]Source: http://www.motohealth.com

developing world to strengthen health systems."[7] The partnership consists of the GSM Association's Development Fund, the U.S. President's Emergency Plan for AIDS Relief (PEPFAR), Accenture Development Partnerships, Motorola, MTN, and Voxiva.

8.1.4 Mobile Marketing

It has long been recognized that the effectiveness of any marketing effort increases dramatically the more it can offer the right product to the right customer at the right time. For example, targeted ads on average generate around 20% response rate while untargeted ads generate around 2–3% response. Therefore, the more is known about the customers and their individual profile information and history, and the more is known about their immediate context, namely, what they are likely to need or want at a particular time, the more effective marketing to that consumer is going to be. Mobile devices and mobile service offer the possibility to:

- Build a rich model of the customer, including behaviors, preferences, interests, social circles, etc. The information needed to create and refine the model can be gathered through monitoring activities as well as through direct input and feedback from the customer.
- Obtain real-time context information about the customer. The timelines of the information is key to the ability to offer the right service at the right time when it will be the most relevant and useful to the customer.
- Quickly access a social network and create viral or buzz marketing [HUG].

As the last bullet indicates, in addition to the value of mobile services to offer a platform for targeted marketing based on customer individual behaviors and preference, the possibility opens up now to an enhanced "social" marketing where members of a social network behave as marketing agents on behalf of the business.

The power of the mobile device to enhance the targeting precision of direct marketing campaigns can be increased significantly if combined with the power of the social network. This enhancement is what is known as "viral marketing" [RAY], where the goal is to "infect" a "susceptible" user, and hope that the user will infect other susceptible users (i.e., spread the word). In this case, the customer's pre-existing social network is used to produce increases in brand awareness or other marketing objective through a self-replicating viral process. It encourages people to pass along a marketing message voluntarily, mostly through "word-of-mouth" delivered or enhanced by the network effect of the Web [BEN]. The basic form of viral marketing is not infinitely sustainable: it is claimed that a customer tells an average of three people about a product or service he/she likes, and eleven people about a product or service which he/she did not like [SIL].

Viral marketing is based on the natural human tendency to communicate. A specific type of viral marketing called "influence marketing" where focus is placed on specific key individuals (or types of individual) rather than the target market as a whole.

[7] http://www.gsmworld.com/news/press_2007/press07_20.shtml

It identifies the individuals that have influence over potential buyers, and orients marketing activities around these influencers. The goal of businesses interested in creating successful influence viral marketing programs is to identify the influential "hubs" which, as mentioned above, are individuals with high "social networking potential," also termed Alpha Users. By creating viral messages that appeal to a segment of the population and by targeting the hubs of that segment, the messages have a high probability of being passed along. "Undercover," "buzz," or "stealth" marketing [HUG] is a type of marketing in which consumers do not realize they are being marketed to. For example, a marketing company might pay an actor or socially adept person to use a certain product visibly and convincingly in locations where target consumers congregate. These "buzz" campaigns can reach consumers isolated from all other media and are magnified by the plethora of communications channels enabled by today's mobile devices. In addition, unlike conventional media, consumers tend to trust buzz marketing.

Since the mobile device is a very personal device, marketing and advertising activities that are conducted through it can be highly targeted and personalized to the user context and preferences. Various technology and services platforms are being used to connect advertisers with end users, through content providers in a model very similar to that employed in Web advertising. Early results that looked at the effect of smart phones such as the iPhone on the effectiveness of mobile ads concluded that there is no particular advantage to sending an ad to a smart phone vs. the older phones.[8] The AdMob platform,[9] which is one of the first successful mobile advertising platforms, connects people who want to buy ads with people who want to monetize their mobile traffic. Through AdMob Marketplace, they offer set of tools for advertisers and publisher, including accurate matching capabilities that ensure optimized targeting of ads to the right users at the right time and analytics for better insights into the effectiveness of an ad campaign. JumpTap and Medio are startups with similar aims and architectures.

The flip side of highly personalized services is the fear and perception of privacy violation and spam. As a society, we are working to address this problem in a responsible way, although it is clear that a comprehensive and permanent solution will probably never be available. One thing is clear though, businesses should inform their consumers on how their information is to be stored, secured, and used or combined with other purposes of marketing [BRU]. If mobile marketing can assist people in their everyday life, they will be more than happy to reveal information about themselves such as their current location. Mobile marketing, like other forms of personalized marketing needs to be permission based and businesses must take great strides in protecting the privacy and respect in the preferences of each and every consumer. Therefore, the interaction must be straightforward and simple. Businesses must not only develop relevant and engaging content with real value that mobile users want to access but must make sure that they follow strict opt-in policies resulting in user

[8]See for example IDV report in: http://www.adotas.com/tag/idc/
[9]See www.admob.com

having choice, control, and confidentiality while ensuring that they only received relevant information.

Although the area of mobile marketing is still in its infancy, it holds great potential, especially if privacy and security concerns are addressed responsibly. As the trend continues, we anticipate the emergence of a new highly personalized marketing paradigm where the message fits the user and comes just at the right time.

8.1.5 Mobile Virtual Worlds

Today, the term "Virtual World" is used in a variety of ways, ranging from computer representations of physical environments, such as office buildings, to computer-based simulated three-dimensional spaces inhabited by *avatars* (virtual representations of users), to a combination of the two. No matter which interpretation one uses, virtual worlds present a great opportunity for research, education, social networking, commerce, and business and consumer applications, to just name a few. For instance, businesses can organize virtual events that can bring together a large number of participants and keep them engaged at a fraction of the cost of a real event.

Perhaps, the oldest virtual world is the Multi-User Dungeon (MUD) game developed at Essex University, England, in 1978 [BARL]. Typical MUDs represent a fictional world where users interact with each other and the world via textual commands. Modern time virtual worlds are mostly represented by Massively Multiplayer Online Role Playing Games (MMORPGs), such as World of Warcraft, and Internet-based sites with user generated content, shared spaces, interactivity, and 2D or 3D GUIs, such as those represented by Second Life, There, and Lively.

Virtual worlds follow a client-server architecture where parts of the world are hosted on a server cluster. Multiple server clusters are used for making up the entire world. Clients interacting with the virtual world are interacting with the server cluster that runs the part of the virtual world they are currently visiting. For instance, Second Life splits up the Earth-like world into square areas of land, called Regions. Each Region is simulated by a single named server instance. While most virtual world clients require custom software to be installed on user computers, recent offering allow Web-based interactions with virtual worlds.

While traditional virtual world clients run on desktop and laptop computers, making virtual worlds accessible from mobile devices has been gaining a lot of momentum lately. There are two main reasons for this. Firstly, the number of mobile devices outnumbers that of desktops and laptops, and this gap is increasing at a fast pace. Secondly, mobile devices are uniquely positioned to bridge the real world with the virtual world due to being always on and connected. Some of the existing offerings in the mobile space are presented in Table 8-1.

A compelling trial called *InsideOut* was staged in 2007 (now ended) by the telecom operator Vodaphone. In this trial, Vodaphone offered its customers the ability to call and text between the virtual and real worlds (the virtual world application was Second Life). At its heart, from the operator's point of view, this was a communication service which used the network infrastructure to enable and charge for a new kind of communications. From the end users point of view, the service bridged the

Table 8-1: Companies and products related to mobile virtual worlds

Company	Description, Distinguishing Characteristics
Comverse	Introduced a Second Life client for Apple's iPhone in 2008. The client software allows users to communicate with Second Life residents using IM, SMS, or MMS, as well as stream real-life mobile video directly into Second Life
Aqros	Offers a thin client that is installed on mobile phones and allows users to connect to Second Life and communicate with friends and neighbors and teleport to different landmarks
Vollee	Originally founded in Israel under the name Game Stream, relies on the available bandwidth in 3G and Wi-Fi networks to offer advanced games and access to Second Life via their interactive streaming video platform
MiniFriday	A research project on mobile worlds developed by Sulake, an online entertainment company that focuses on virtual worlds and social networking and the creator of the teen-focused virtual world and social community Habbo. A simple world consisting of a bar and a club with different languages. The client application is only available on Nokia S60 handsets
Lamity	Selected by NTT DoCoMo to be bundled on the latest 906i line of phones from hardware manufactures Sony, Sharp, and NEC in 2008; a Java-based application that utilizes AJAX for sifting the processing from the mobile handset to a game server

virtual and real worlds by providing a convenient and fun way to talk between them (Fig. 8-1). A simple scenario for calling *into* Second Life was as follows: Tim knows that his friend John is currently playing Second Life (Tim is not) and wants to talk with him. Tim uses his Vodaphone cellular phone, dials a special number, and moments later John's avatar's virtual heads-up display rings with a call. As John commands his Second Life avatar to pick up, a communications

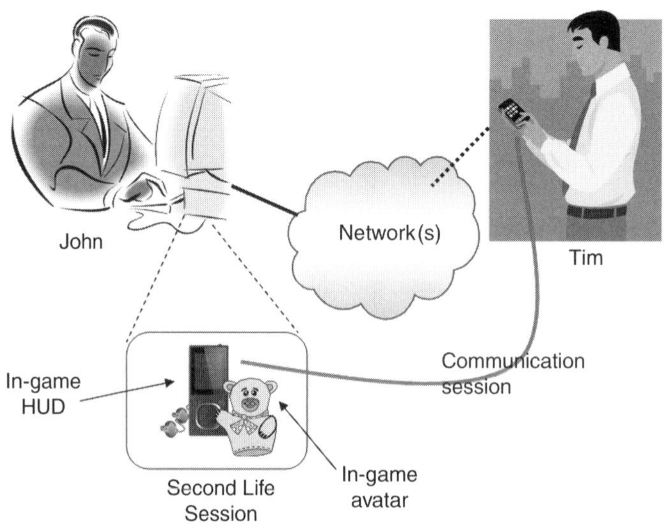

Figure 8-1: High level schematic of the Vodafone Second Life initiative.

session begins that is anchored at on one side at Tim's real cellular phone and on the other side at John's computer. Between the two endpoints was Vodaphone network elements and Voice-over-IP (VoIP) architecture. The other main use cases include: a) John's avatar in Second Life calling out of the game to Tim's cellular phone (John can then use the VoIP client embedded in the game to have a voice session with Tim), and b) John on one Second Life island calling a player on another island (e.g., in-game to in-game call). Equipment was needed on both the real and virtual sides. On the virtual side, the virtual equipment that avatars need to take part in voice sessions was called the Vodaphone "Heads Up Display" (HUD) and could be "purchased" at a particular Second Life island from a vending machine. On the real side, cellular users had to do little more than register their compliant cellular phones as well as some other administrative procedures. This service's current status is dormant, despite the mostly positive enthusiasm from the technical pundits (many of whom compared it to a main gimmick in the storyline of the movie *The Matrix*). One can only expect more such trials and experimentations as developers and users get familiar with the new kinds of multi-modal interactions.

8.2 ADVANCED AND EMERGING MOBILE TECHNOLOGIES AND SERVICES

This section outlines some innovative services that make use of advanced capabilities, either in the network, in the mobile device, or at some conceptual level. Mobile device and services research is currently experiencing a sizable renaissance. The rise of open handsets and Integrated Development Environments for device-side application development does not hurt, nor do the increasingly capable devices that are more and more affordable to universities and graduate students, small laboratories, and individuals. A wide swath of mobile technologies and services can be built quickly and relatively cheaply, and many players are getting into the game. These and other technologies are enticing to some companies who believe that new and innovative ways of reaching customers will eventually reap rewards. In the rest of this chapter we take a rapid-fire look at technologies such as:

- New business-to-consumer "auctioning" models based on user context;
- Mobile devices and their capabilities to play new roles such as scanning and the push towards repurposing;
- The rise of ambient and ubiquitous mobile services, as well as innovative use of augmenting reality via the mobile device;
- Open APIs for mobile service development and some of the highly innovative services that have resulted from developers;
- New device and interface types that "bend the rules" of today's rigid equipment.

These descriptions should both serve as starting points for the reader as well as provide an excellent broad view of today's and tomorrow's innovation.

8.2.1 Auctioning Models Based on User Context

Mobile commerce, in which on-the-go consumers transact business through their mobile devices, is already here, to a large extent. A big driver of m-commerce on a per-user scale will be mobile advertising. Google and Overture/Yahoo! are two of the biggest players in this regard. Besides advertising, a highly compelling scenario is emerging in which a location-sensitive auction takes place in which auction participants submit bids for services in an attempt to win a market-maker's business. In other words, a mobile customer (market maker) is driving through a town at around noon. The customer is hungry and informs the service as such. The service responds by beginning a real-time auction amongst the local restaurants in the town in which the provider with the "best offered deal" for the customer will win the right to communicate directly with the customer.

Implementing and supporting mobile Dutch reverse auctions may require different procedures than that of English auctions.[10] Dutch auctions may be concluded by the submission of the first bid. For example, in a Dutch auction the market-maker would announce a set price (often low) to the candidate affiliates, and the first affiliate to respond would win the business. In this case the auction execution requires reliable and timely delivery of bidding information. Therefore the platform should ensure that all communications involving bids are delivered with the highest degree of reliability and timeliness. Dutch reverse auctions also do not require that auction participants be made aware of other participant's bids. The Dutch reverse auction can be thought of as a yes or no proposition put to each of the candidate affiliates. The first to answer yes wins the business. In this case, the non-winning participants only need to be notified that they did not win. One implementation of a Dutch system would close the bidding process once an affiliate accepted a Dutch reverse auction. In this form, when a subsequent affiliate was directed to the bidding Web-portal, they would be presented with a notice that bidding was closed. In another form the system allows the system to prioritize the candidate affiliates and determine which subset will be contacted first, and thus have an advantageous bidding position. This capability could be used by the market-maker to tier their local service provider affiliates, extract additional revenue or encourage local affiliate behavior. There are several steps that might help enable market-makers to successfully and reliably offer real-time reverse auctions. They are outlined below.

1. Geographic localization and confirmation of customer's location;
2. Categorization and indexing of customer requirements;
3. Possible access to customer data;
4. Identification of feasible local service affiliates;
5. Screening of feasible service affiliates to candidate service affiliates;

[10]Auctions may be considered English or Dutch depending upon the nature of the starting price set by the market maker. In English reverse auctions, the starting bid is the highest price the market-maker is willing pay for the service, and subsequent prices are lowered. In Dutch reverse auctions, the starting price set by the market maker is low and subsequent prices are increased until an auction participant agrees to pay it.

6. Notification of candidate service affiliates;
7. Display of relevant customer request information to candidate service affiliates
   a. Initial price for bids
   b. Distance to customer and distance to desired customer location
   c. Real-time traffic and weather conditions for both the customer's location as well as the customer's desired drop off point
   d. Routing information from the local service affiliate's location to the customer, from the customer's location to the customer's drop-off point, and from the drop-off point back to the service affiliate's location
   e. Applicable history and profile information for the customer
   f. Make and model of car, as well as possible items to carry to the scene
8. Auction execution software
   a. Mechanisms that allow the candidate service affiliates to bid
   b. Real-time data management methods that can reliably rely bid prices from auction participants to the auction platform
9. Settlement software that finalizes a contract between the winning bidder and the market-maker;
10. Billing mechanism that allows the winning candidate affiliate and the market maker to be paid;
11. Feedback mechanisms to allow both the customer and the service affiliate to submit feedback about the process that will be stored in a database and consulted for future service opportunities.

In summary, all of the above steps and procedures require careful study and implementation when being supported by a market-maker and offered to mobile customers. Business-to-consumer use cases might also include car repair stations bidding to win a customer whose car is broken down nearby, or hotels bidding to win tired travelers at day's end. Consumer-to-consumer use cases might also be compelling (see Chapter 6 for further details).

8.2.2 Repurposing the Mobile Device

As the mobile communication devices become ever more capable and ubiquitous, a new style of usage is emerging in which researchers and hobbyists use the mobile device in novel or unexpected ways such that new forms of interactions or services are possible. Let us begin "at the end," with some simple examples. It is now well-known that the accumulation of expired and trashed electronics is a possible health and environmental concern (persistent bioaccumulative and toxic chemicals are the main concern: arsenic, antimony, beryllium, cadmium, copper, lead, nickel, zinc, etc.) The ultimate end-of-life repurposing for a mobile device is to relay it to an organization such as wirelessfoundation.org or recycleforbreastcancer.org. These organizations accept technologies that are being trashed, salvage parts, and sell them to refurbishers who get them back onto the market. In cases where nothing can be refurbished there is still an incentive for using these sorts of services as many of them will ensure that the device is recycled as safely and in an environmentally friendly fashion.

For do-it-yourselfers with modest know-how, modern mobile handset at the end of their usefulness as phones can be "downgraded" single-function devices and passed on to people who need them. Viable recipients might be children (toys) or needy individuals or charitable organizations. Of course a hand-me-down still has to eventually be recycled but at least the life is extended. Examples of downgraded devices include:

- Using the phone as solely a music player, or as a document-backup system;
- Using the phone as a portable digital camera;
- Using the phone as a calculator, note taker, or calendar only system.

In another wonderful example, Rubinsky, a researcher at UC Berkeley has developed another sort of repurposing that fills a gap in medical imaging [YAN]. Rubinsky realized that access to ultrasound, MRI, and X-rays in development countries was (and is) very limited, yet, on the other hand, it is well-understood that medical imaging is a cornerstone of modern medicine. Its use in disease diagnostics, tumor and tuberculosis detection, and fetus monitoring, is essential. The World Health Organization estimates that about three-quarters of the world's population has no access to medical imaging. Rubinsky felt that developing a repurposed agile approach was more practical, quicker, easier to maintain, and cheaper than the option of donating expensive imaging equipment to these countries.

His innovative solution was to physically separate the two main functions of imaging (normally embodied in the same hardware): data acquisition and image processing and presentation. His prototype allows for data acquisition tools to acquire and upload imaging information to a central server by using a cellular phone that is attached to a network. As the raw imagery data hits a central server, it is processed using sophisticated and time-intensive computations. When an image results from the computations, the network can again be used to retransmit an appropriate image back to the sending cellular device where it can be inspected and analyzed by a health professional. Electrical Impedance Tomography (EIT) has been used and relatively affordable acquisition devices have been used – in one example, the EIT data sent up to the server is only about 6 kb (a series of voltage measurements through tissues), but the resulting imagery shows sufficient data to inform a health provider (Fig. 8-2). Rubinsky's work is provocative, and makes one wonder about other uses in other realms. What about:

- Using the server-side to compute a social network analysis for a given user and emitting to the user's mobile device screen by way of a compact and effective visualization method (e.g., social network map, etc.)?
- Using the server side to execute network analytics and to render graphics for a network technician who is servicing switching centers. The graphics (maps, schematics) are rendered on the technician's phone?

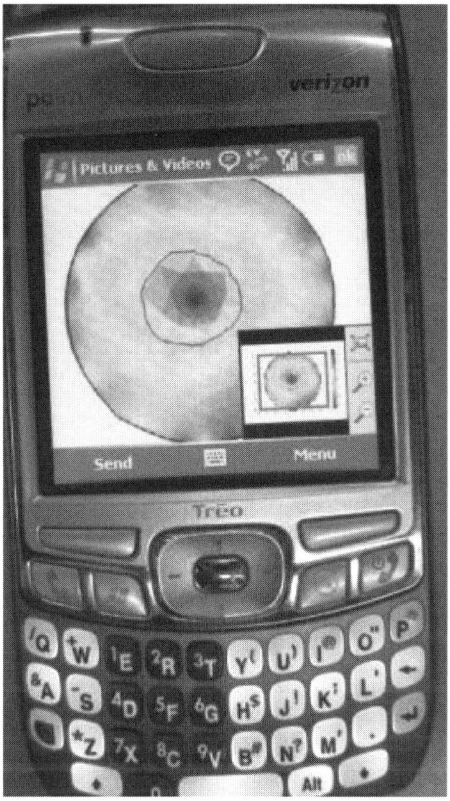

Figure 8-2: Rubinsky's use of cell phone technology allows data acquisition and image analysis steps to occur on different platforms – notably, in developing nations, the latter can occur on cheap mobile devices. An emulated breast tumor is shown here. Used with permission from Boris Rubinsky.

The technique is reminiscent of a similarly-spirited product called Nike + iPod, by Nike and Apple. In this well-known, mass-market application, a pedometer communicates with an iPod attachment that in turn brings the user diagnostics, information, and music relating to the current walking or running session.

In another section of this chapter we describe the Android handset Developer Challenge in which the developer community was invited to create interesting applications for the Android operating system (with $50 million on the line in prizes). Because the developer tools allow the programmer to access and repurpose onboard device I/O units, several applications emerged that can be thought of as phone repurposing. As an example, one team developed an application that effectively transformed and Android phone into a super-deluxe barcode reader. The device could then be used to scan barcodes of any product and – after a backend search – to display community notes, reviews, and merchants that relate to the product. In another example, the phone is transformed into a musical instrument. The *Rayfarla* application repurposed the touch-screen of the device into a touchable musical keyboard. By changing simple settings such as tone (e.g., organ, piano, bass) and then tapping this keyboard the user can play a simple song. The application also presents a series of mini-games that revolve around pressing musical keys, identifying music, and so on. In *SplashPlay*, the Android handset is transformed into a musical learning device. When teamed with an external Bluetooth-enabled Light Panel that gets attached to the head of the guitar, the handset becomes a guitar teaching station. Songs played on the device-based application are beamed to the Light Panel, who in turn, by beaming light onto the first five frets of the guitar, illuminates the frets that should be pressed to make the notes of the song. All this happens in real time. Songs and lessons can be downloaded and shared and many aspects of the data are open source and community based.

8.2.3 Ambient Services and Interactions

Ambient displays are a relatively novel (at least for the mass market, if not the research) new way to display information. Ambient displays convey information in calm, environmental ways that do not demand the full spectrum of user-attention. In practice, examples of ambient notifications have come from a wide spectrum of modes:

- A glowing desktop orb, changing colors in response to energy usage in the household;
- A vibrating phone vibrating in response to new information;
- Furniture and equipment – such as water fountains, lights, flowers, and mobiles – that subtlety change to indicate information changes.

Ambient systems and delivery modes are particularly suited for situations in which users have greatly divided attention and are multitasking. On-the-go mobile users usually meet these criteria. They often rely upon and assume that the user's periphery vision will notice the changes made programmatically by the system and take appropriate action. They most often deal with conveying noncritical information to users and assume that critical information will be conveyed *directly* into the user's field of view or attentional range. [POU] defines a useful taxonomy that classifies such systems based upon the following four characteristics:

- *Notification Level* – This is considered the level of ambience of the system, or the degree to which the system is able to (and meant to) be intrusive to capture the user's attention. A glowing orb has a low value while an alert that pops up directly in the users attentional is high.
- *Representational Fidelity* – Pertains to the sophistication of information that can be displayed using the ambient method, and how effectively the system uses: symbols, letters, drawings, lights, maps, colors, and so on, to convey the information.
- *Aesthetic Emphasis* – Relates to the success of the ambient system as art in and of itself and is a fairly subjective – yet intuitive – measure. Many systems use

artistic paradigms and industrial and interior design tricks and patterns in order to make their ambient products more "invisible," more compatible with the décor of modern rooms, or more like some common object.

- *Information Capacity* – Relates to the number of separate information pieces that the ambient system is able to handle. Some systems display only the user's energy consumption level as a color on an orb (e.g., have one dimension), while others use various rotation techniques and multi-level display to show up to 20 different information dimensions. There is clearly a tradeoff between the information capacity and the simplicity of device.

Philips Research is only one lab that is interested in the commercial potential of ambient like systems. Their vision of ambient computing includes, "people living easily in digital environments in which the electronics are sensitive to people's needs, personalized to their requirements, anticipatory of their behavior and responsive to their presence." Their *Ambilight* system is a series of lights that enhances television viewing by illuminating in particular combinations. For example the system may use the lights to wash the room in a blue hue during an underwater scene. Ambient Devices, Inc. – an MIT Media Lab spinoff – has marketed several ambient style products to the mass market, including: umbrellas, orbs, dashboards, forecasters, and clocks.

What about ambient apps for the mobile device? There is little research aimed directly at ambient notification using only the mobile handset. This is largely because, up until recently, the mobile handset resources that were programmable and available as I/O for ambient notifications were limited to: the screen, the vibrate mode, and the ringer. Using the screen alone for such notifications may be a failure since when the user's attention is not directly upon it, there is no information exchange. The use of vibrate and custom ring modes is promising. With the rise of open handset operating systems and ever more advanced technologies (see section on flexible and organic interfaces) there is now much more promise that ambient notifications will bleed into the mass market through things such as wearable fabrics, sunglasses, touch-sensitive morphing devices, and so on. Certainly, given its connectivity to the infrastructure and Internet, the mobile handset is – and will remain – a key part of the full ambient information system use case. Even if the notifications are delivered through other sorts of technologies the mobile handset embodies the radio network and will therefore provide an important link to the greater Internet, needed for metadata and customized information.

A schematic of the components comprising a hypothetical networked ambient system are shown in Figure 8-3. In this figure the user's location is resolved through positioning (e.g., GPS) and location services (e.g., reverse geocoding and so on). Once positioned an Ambience Service component can trigger changes to the environment near the user, such as to connected furniture, lighting, digital systems, networks, and storefronts. For example, when walking down a street of storefronts, the storefronts may change their lighting in concert with the user's progress down the street – perhaps in accordance with the user's "favorite" colors or some other policy stored in the Ambience Service [DAN].

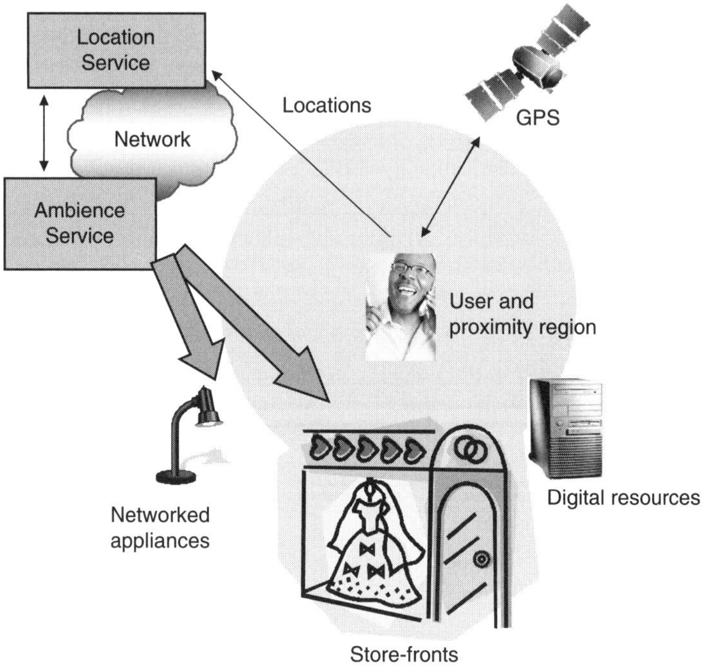

Figure 8-3: Actors in a networked ambient system.

8.2.4 Google Top 50 Android Applications

In 2007, coincident with the release of the software development kit for the Google Android project, Google announced the "10 million dollar Android Developer Challenge." Android, "a software stack for mobile devices that includes an operating system, middleware and key applications" entered the mobile scene with much fanfare, touting an open environment that leveled the field for developers and let even the basement-developer dream that her application might be available on Android hardware.

A main reason the excitement level is so high is that the Android developer has programmatic access to many system level functions that she simply does not in other mobile operating systems. As a result, applications have the potential to be richer and more diverse, interworking with onboard calendars, incoming text messages, timers, XML/HTTP servers, Google maps, onboard I/O and cameras, and so on. In mid-2008, after all the Developer Challenge entries were judged on the basis of: originality of concept, effective use of the Android platform, polish and appeal, and indispensability, a list of the top 50 applications was released. These developers get a chance to continue into phase 2 of the contest where hundreds of thousands of dollars may be available to each of the top 10 or so. Many of these top 50 applications reflected an infatuation with location and feature location-sensitive information or maps with informational overlays. However, many of the developers also provided interesting, if not modest, solutions to real world issues and concerns such as biometric identification and personal safety.

The following summarizes some of the more interesting top 50 Android applications[11]:

- *CookingCapsules* is an application that is useful in two main contexts: the grocery store, and the kitchen. To begin, the user watches a brief (two minutes or so) Internet video cooking demonstrations. If they are interested in the recipe, the application develops a list of ingredients and a shopping list that completes the recipe components. The application then can use a directory of services to locate and direct the user to a nearby "market" and can also send the shopping list (via email or SMS) to a friend or spouse for pickup. Having chosen a recipe and now with ingredients on-hand, the application displays a clean list of steps that provide guidance through the recipe and that can be "checked-off" after completion.
- *AndroidScan* is an application that allows the user to employ her Android-based handset as a barcode scanner. The development team employed open-source scanning software called Zxing (see http://code.google.com/p/zxing). Zxing is an "open-source, multi-format 1D/2D barcode reader library implemented in Java ... to support decoding of QR Codes, Data Matrix, and the UPC family of 1D barcodes." AndroidScan has many use cases but primarily: 1) the user scans the barcode on the back of any product with a barcode (CDs, books, etc.). The application then does a Web search to find and present a) details about the product (including reviews, ratings, etc.), and b) merchants that offer the product for sale and their asking prices. For books, the application can search libraries near the user's location by first reading the GPS coordinates of the handset and then issuing a search across libraries that are "nearby." Direct and immediate access and exploitation of Android's (and Zxing's) scanning capabilities is the key new feature here; a clear and simple user experience is also a nice differentiator. Other details about mobile bar-codes are presented later in this section.
- *CallaCab* is an application that helps the mobile user call a cab for pickup at the place where she is currently standing or at some future time. The application links with onboard GPS to determine the user's location, uses maps to overlay cab locations and does some interpretation, making it possible to call and reserve cabs when travelling through almost any country. The application interworks with the telephony API in order to make calls. *CallaCab* is simple, but is one of those "ah-hah" applications that, by this time, one would think would be preloaded onto all new cellular phones (especially for customers in urban areas).
- *TuneWiki* is an application that attempts to display lyrics in synchronization with song playback on the device. First it indexes the songs on the device and presents them in an easy list. When any indexed song is subsequently launched by the

[11]Complete list and more definitive details can be found at http://code.google.com/android/adc.html

user, the application communicates with a back-end server and attempts to download song lyrics and album art (via other third party services). During song playback the song's lyrics are displayed in rough synchrony with the music. If the synchrony is not perfect, the user can restart the song in "tap mode," listen to the song, and tap on the screen each time the displayed lyrics should advance to the next line. Subsequent plays of the song will retain this improved synchronization.

- Repurposing the device to become a music player or instructor – Elsewhere in this chapter we have described two Android applications called *SplashPlay* and *Rayfarla* which are novel applications of touch screen capabilities.

In summary, it is clear to see that the fabric of mobile services, from the point of view of the end user, is truly being "re-weaved" by the changes in the developer landscape and device openness. In a matter of months, developers have been able to create interesting and useful new applications for Android, a process that was facilitated by Eclipse-based development kits and emulators. In the past, locked devices and inconvenient roadblocks (vis a vis certifications and so on) to application integration on the network have, in some sense, been barriers to innovation. Linux, Android, and other open handsets are changing this [LAW]. It should be noted however, that the top 50 applications can be thought of as falling into two categories:

1. Those that assume, rely or exploit positioning (and, almost always, maps) and telecommunications use cases.
2. Those that are novel but do not use communications in any non-trivial ways.

The latter types are interesting but do not define the fabric of new mobile services so much as the former. Furthermore, while novel, we do not claim that each of these applications is bound to generate millions in revenue. Indeed, since most were written by individual users in informal ad hoc fashion, the formalities of user studies and human factors may have been glossed over in favor of flashy interesting features. Discovering exactly what mobile customers want will continue to be a challenge for mobile operators before, and now after Android's appearance.

8.2.5 Mobile Augmented Reality

Recent research efforts by industrial research labs and universities focus on taking advantage of the wide-spread adoption of cellular phones with embedded cameras in order to enable accessing information about a particular object (e.g., landmark, restaurant, person) from the Internet by capturing the object using the camera and then using the captured picture as the input to a search engine. The search engine results are typically displayed on top of the image on the phone's display, as shown in Figure 8-4.

Being able to retrieve information about a particular object from a mobile device with the least amount of effort is clearly a challenging problem. While using a traditional search engine may work when certain properties of the object in question

Figure 8-4: The Nokia Mobile Augmented Reality Applications (MARA) poster. Reproduced with permission of Nokia.

are known (e.g., street address, business category) and used as the search keywords, in most cases this approach is either impractical or requires extensive user input and search query refinement. Taking advantage of cellular phone capabilities, such as embedded cameras and various types of receivers and sensors, presents itself as a more efficient and user-friendly way to address this problem. Several prototypes are available in this space. Bill Gates demonstrated Microsoft's research efforts in this space during his last Consumer Electronics keynote speech in 2008.

Nokia's Mobile Augmented Reality Applications (MARA) prototype utilizes camera equipped mobile devices for video see-through mobile augmented reality. The MARA prototype annotates the user's surroundings that are within the focus of the camera by overlaying the image stream captured by the mobile camera with graphics and text in real-time, illustrated in Figure 8-4. The MARA prototype depends on additional hardware components beyond the embedded camera in order to determine the parts of a scene that is viewed by the camera. In particular, it requires a GPS receiver for determining the position of the mobile device, accelerometers for determining the orientation, and a compass for determining heading.

8.2.6 Lifestreaming

Microblogging services allow subscribers to post short personal messages online or broadcast them as text messages. Since the launch of Twitter in July 2006, the first major microblogging service, an entire ecosystem has been developed around microblogging services. A recent high-impact trend that extends the message-oriented nature of microblogging is lifestreaming. Lifestreaming aggregates your recently created information and media into a single steam, providing a timeline of your recent activities. For example, with lifestreaming a posting on Twitter, an instant message exchange with a friend, a video upload on YouTube, and an image upload on Flickr could be aggregated into a single stream of messages and content.

Lifestreaming presents an enormous potential for personalized and social mobile services. By using the image and video capture capabilities of today's mobile phones, coupled with automated or manual attachment of location information to captured media, almost everything one does could be logged and tracked. Nokia has already jumped into this trend and will be making available a lifestreaming application named LifeviNe in the fourth quarter or 2008. LifeviNe allows geo-tagging of captured content and manual or automatic upload to a Web site where it can be further tagged and shared with other.

8.2.7 Evolution of the Device: Smart and Organic

Not too long ago, mobile phones were voice and SMS-centric, did not include any smart operating system, had limited capabilities, small screen size, and were evolving at a relatively long life-cycle (more than 18 months). These phones were relatively cheap and positioned towards emerging markets, first time users, and pre-paid solutions. Their limited functionality restricted the supported mobile services to SMS-based informational messages (e.g., reminders and alerts), DTMF and speech-recognition based voice portals, and simple Web interactions via the Wireless Access Protocol (WAP) for WAP-enabled phones.

Basic mobile phones evolved into multimedia-enabled phones offering large colorful displays, integrated digital cameras, MP3 players, Internet browsing, email synchronization, and Instant Messaging functionality. While these phones had high prices attached to them during their initial introduction, increased competition and carrier subsidies have reduced their prices considerably (even made them free) in the last couple of years. The advanced functionality and relatively cheap prices of multimedia phones make them good candidates for many mobile services such as multimedia content access (e.g., ringtone and wallpaper downloads), mobile commerce (e.g., mobile banking and ticketing), social networking (e.g., Nokia's Ovi), and mobile games.

Wireless Personal Digital Assistants (PDAs) and Smartphones represent the most recent iteration of the mobile phone evolution – data-centric mobile terminals that include a QWERTY keyboard and offer wireless email and calendar synchronization. Smartphones, on the other hand, are voice-centric and utilize an operating system for running software applications. It is now not easy to tell one from the other (just line up BlackBerry's Curve, Apple's iPhone, Samsung's Instinct, and Palm's Treo and you'll see!).

Nowadays, the seemingly never-ending advances in materials, storage, and even nanotechnologies, are leading to devices and paradigms that will change the way we think and interact with mobile services and information. Electrophoretic ink, light-emitting polymers, thin-film circuits, nanotech-based fiber mesh, transparent electronics, shape-memory polymers, ultrasound motors, and quick-charge batteries are being embodied in today's conceptual prototype devices, but promise to be integral day-to-day parts of tomorrow's mass market technology. Organic user interfaces are *distinguishable* from standard interface technology on three main criteria: 1) the input components of the devices also tend to be the output components; no distinction between the two is necessary, or designed in by the designers. 2) The shape of the device, much more than with non-organic UIs, defines its functioning. For example, a rounded torus-like touch device will be intended to be worn as a bracelet by the user; a icosohedral display's interface will deftly take advantage of – and benefit from – the device's shape by, for example, showing pertinent information on each side, possibly exploiting the natural adjacency between sides of the shape to imply relationships. 3) The shape of the display is not fixed or static [VER]. New technologies already allow folding screens and bendable shape; the form of the interface will, at best, "follow the flow of user interactions" [VER]. The act of interface-reconfiguration may also become a part of the interaction use case with the user; that is, each reconfiguration may come to mean something. For example, a bendable torus-shaped interface may automatically begin to stretch in the user's hand, to indicate that some important information is to be conveyed.

Elsewhere, a compelling prototype called *LucidTouch* (Fig. 8-5) has emerged from the joint labs of Microsoft, University of Toronto, and Mitsubishi [WIG]. LucidTouch addresses two key problems with touch screen devices, such as the iPhone:

- *Occlusion Problem* – the display is occluded by the finger in the key moment just before touch.
- *Finger Suitability* – the touchable size of a person's finger is much larger than pixel dimensions, making interactions difficult.

Past research initiatives have indeed placed touch areas on the back side of devices, but only with limited success, partially due to the fact that one could not see where one's finger is with respect to the screen (e.g., the device itself occludes the finger). The LucidTouch solves this by overlaying an image of the users hand and fingers onto the front of the screen, giving the appearance of transparency. The image is created in real-time, in the prototype, by performing various image processing techniques on video imagery captured by a camera attached to the device itself. Tracking feedback to the user is given in a variety of interesting ways, including rendering colored dots on the images of the behind-the-device fingertips in accordance to whether or not the fingertips are in contact with the screen, and so on. Thus, the device's effect is a lot like a piece of frosted glass with the user's hand partially in front and partially wrapped around back of the glass – the fingers on the backside are shown graphically on the

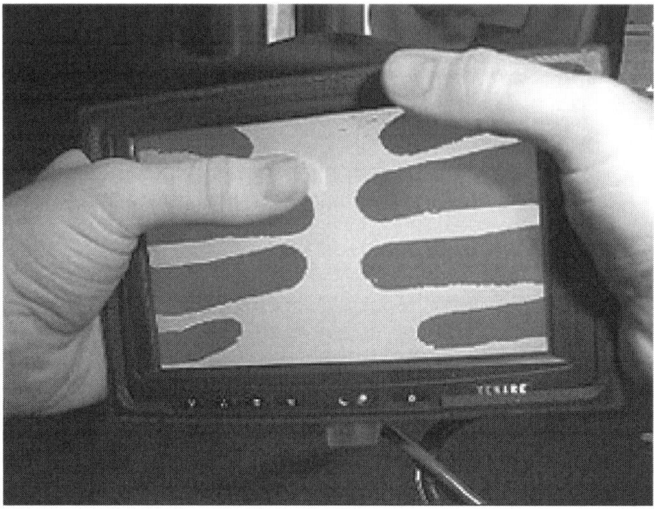

Figure 8-5: Microsoft LucidTouch device (Source: http://research.microsoft.com/ users/ baudisch/projects/lucidtouch). Used with permission from Patrick Baudisch.

front side, complete with tracking effects. The device, therefore, allows many novel interaction techniques that combine touches on both sides of the screen, such as soft keyboard interaction and map browsing interactions.

Just as every wave of new device types requires end users to learn new ways of interaction, they also require service developers to become adept at creating services that allow users to achieve the main user tasks (see Chapter 4 on Simplicity). Using a complex new device for a task that does not merit it, or imposing changes on long-standing interface metaphors with which users are comfortable, can create usability, technological, and experiential roadblocks. Early adopters of these devices often pay a high price for the hardware, as well as experiencing a dearth of software and services that work well with the device. This usually lasts until either there is increasing supplies and demand for the product and when the development community becomes large and experienced.

8.2.8 Mobile Barcodes

The popularity of the original barcodes (vertical black lines and white spaces of variable width used for encoding numbers) and the desire to encode more than just numbers led to the development of a new generation of barcodes that can encode thousands of alphanumeric and other characters in virtually any language. These barcodes, referred to as 2D, matrix, or mobile codes, are formed by patterns of black and white shapes (typically squares) arranged on a grid. Figure 8-6 shows an example of such barcodes containing the text: "*The Fabric of Mobile Services*."

While this new generation of barcodes was developed in the 1990s, their enormous potential became apparent when barcode reader technology allowed the embedded camera in mobile phones to act as a reader. The most common use of mobile barcodes

Bee Tagg **QR-Code** **Datamatrix**

Figure 8-6: "The Fabric of Mobile Services" barcode generated using the Bee Tagg Multicode generator.

is to provide information about a specific item. While some of this information may be directly encoded in the barcode itself, in most cases the information is accessed from a Web site by visiting the URL encoded in the barcode. It is this Web-based link that makes mobile barcodes extremely powerful, especially for marketing, loyalty, advertising, and entertainment purposes. For example, the *Sun* newspaper in the United Kingdom includes a mobile barcode section that allows readers to go to *Sun's* mobile Web site. In Japan, mobile barcodes have been extremely successful, in part because most of the handsets include a mobile barcode reader, and in part because more than half of the population is accessing the Internet via a mobile phone.

An important property of mobile barcodes is that they require consumers to take action in order to access the information encoded in them. This property allows a permission-based approach to advertisement and marketing and, hence, it eases consumer angst associated with intrusive mobile marketing and advertising campaigns in the form of unsolicited spam via SMS, MMS, or voice calls. Despite their potential and promising success, however, future mobile barcode success depends on several factors. First, many of the mobile barcodes use their own format and reader software, such as QR Codes, Data Matrix, Aztec, Upcode, mCode, Trillcode, Quickmarks, ShotCode, and Beetagg. Without standardization efforts, the mobile barcode market will remain fragmented and negatively impact widespread acceptance. Second, many operators still favor walled gardens and restrict their subscribers from accessing services not offered by them or their partners. Third, mobile barcodes require substantial back-office integration in order to provide mobile barcode transactions that are both accurate and accountable because charges may be applied to these transactions. Finally, readers and back-end services must be secure in order to avoid privacy violations and unauthorized accesses.

8.3 GETTING PERSONAL – WHAT IT TAKES AND WHY IT'S IMPORTANT

In today's mobile world, personalized services are gaining a prominent role due to their potential to cater to the specific needs of their users. A basic element of a personalized service is user-related context information such as policies, preferences, location, presence, mood, schedule, and social networks. While some elements of user context could be captured manually and explicitly, the capture and management of time-depended elements, such as location, presence, and availability, faces many technical, business and psychological barriers. And it is only after this entire context has been gathered and processed that services are able to "get personal." A personal service has the potential for a greater psychological and experiential impact on the end user, and a positive user response can be more easily funneled into revenue. Dawn Nafus has been examining the interplay between collecting data about the user and determining the user real context as well as the relationship between Return on Investment (ROI) and Culture [NAF]. The conclusions are not surprising – sensor data and location information do not provide the actual personal context of the user that includes emotions, moods and culture. However, we can correlate what we can measure with what people value (as reflected in the ROI) which is related to the culture which is how people understand the world around them.

As we have seen throughout the book, mobile services and applications are used in varying contexts and surroundings. When the services are made context-aware, they can offer contextually relevant information to the user. The recognition of context by the system means that the user's information needs can be anticipated and responded to in a more automatic manner. Information and service needs vary according to user's immediate situation (see lengthy discussion in Chapter 6). A context aware service responds to this by offering contextually relevant information. For example, in the case of deep context navigation described in Chapter 6, only messages that might have an impact on the current context are not discarded. Also, contextual information can be shared – to good effect – with other people in the user's social "surroundings," such as in social networking systems.

Personalization and adaptation of services for different users, different contexts of use, and different devices are the most promising factors in matching the service with the user and thus improving the usability of the service. Technical challenges of supporting these services include understanding the context in which mobile handsets are being used, recognizing the strengths and unique challenges of mobile handsets, and providing a simple and intuitive user interface are key factors in successful user adoption, as we discussed at length in Chapter 4. Furthermore, it is well known that simply porting PC-bound desktop applications and services for mobile consumption does not work [STU]; thus, even in the case of existing, highly effective, personalized desktop services, it remains non-trivial to create the mobile offering. On the business front the challenges include developing strategies for enabling advertisers and content providers and making the service available to a wide audience is required for profitable services. On the sociological and psychological front concerns about privacy and sharing of collected information can significantly reduce service up-take; they can be somewhat abated by enforcing strict security and data life-cycle management.

Throughout the book we have seen that location (and other) information is one of the key user context components that become available with the mobile device and can increase the experiential value of mobile services. Furthermore, the convergence of mobility with specialized devices such as electronic book reading devices (e.g., Amazon Kindle), game devices (e.g., Microsoft Xbox), or a vehicle device (a telematics-enabled car), provide additional context as well as additional modes through which the user can be reached. For example, such services might include sending targeted emails to users based on their Kindle book purchases, or to those mobile users that are playing the same game or strolling in the same area at a given moment.

8.3.1 Challenges of Personalized Mobile Services

Personalized mobile services are characterized by the need to manage massive amounts of data. Some parts of the data – such as user preferences – are relatively static, while some other parts – such as user location – are highly dynamic. Furthermore, some portions of the data – such as calendar appointments – are relatively small in size, while some other portions – such as pictures, audio and video – may be quite large. Some of the data may be well structured, such as location information, while another subset of the data may be unstructured, such as SMS and email messages. The ability to manage such large amounts of diverse data in an efficient and intelligent way and provide real-time services presents many technical challenges such as:

- Processing diverse data streams in order to capture real-world events;
- Balancing privacy against the benefit of sharing real-time information;
- Intelligent processing and sharing of information between physical and virtual worlds;
- Development of appropriate storage and retrieval methods, data processing engines, parallel and distributed architectures and software techniques for managing the communications and events across a huge number of users.

Besides managing the large amounts of data associated with mobile personalized services, being able to mine the collected data for marketing and advertisement purposes faces its own unique challenges. While analytics and marketing technologies are well established in the enterprise market and the wired Internet, applying such technologies to mobile services has proven to be quite challenging due to some unique mobile device properties. Two main issues posing challenges include:

- Diverse device capabilities
- Tracking mobile devices

The plethora of mobile devices, each with its own unique capabilities and shortcoming, coupled with the various carriers that serve these devices and the different service plans available to consumers make it extremely difficult to determine which content to deliver and how to deliver it. While having access to a mobile device capability database – such as the DeviceAtlas database available by dotMobi – enables

Web sites to determine the media types a device can handle, being able to determine carrier information and service plan details may be needed for being able to deliver a truly personalized service. In particular, carrier information is valuable because it can be used for determining available connection bandwidth in order to select the correct media format to deliver (carriers often limit the upload/download bandwidth available to consumers). On the other hand, information about the specific service plan details associated with the device can result in reduced costs associated with media accesses. For example, a higher resolution and longer video clip could be selected for an unlimited data plan, whereas a shorter and lower resolution video clip would be offered when data usage charges are applied.

On the topic of tracking of mobile devices, as mobile devices roam from one base station to the next or move from a cellular connection to a Wi-Fi connection (dual-mode phones), their IP address may not remain the same. While the MobileIP standard from the Internet Engineering Task Force (IETF) allows mobile devices to maintain a permanent IP address while roaming, support for this standard is still lacking widespread adoption by carriers. Therefore, being able to tie Web site accesses to a specific user becomes challenging. Aggregate analytics could be collected by using existing Web page tracking techniques, such as those that include a one-pixel transparent image in every Web page.[12]

The increased processing and storage capabilities of current and future mobile devices allow the development of forward-looking location-aware services and middleware components that utilize innovative technology in order to optimally utilize resources in user mobile devices, in other mobile devices (in peer-to-peer environments), in third-party service providers and in telecom network-based services (e.g., IP Multimedia Subsystem). Such services will be using historical data and real-time session information to enable intelligent reminders and alerts, location-based billing and targeted advertizing, and other techniques.

From an architectural viewpoint, most of today's applications and services follow a thin-client thick-server architectural model, in which most processing occurs on the server. JavaScript and AJAX technologies have started to change this model and allow for processing to take place on the client.[13] By architecting mobile applications and services in such a way that they run more effectively, more locally (when applicable), and with more individualized context, perceived user experience is likely to increase. In such a paradigm, processing and resources can be fluid and placeless and a new challenge becomes developing appropriate middleware components that span the mobile device and back-end servers and

[12]In this technique, when the Web page is loaded, the client browser attempts to fetch the image from a Web server which records the IP of the accessing device together with additional information associated with HTTP headers. By doing so, it could generate Web page and content access statistics across the various devices accessing these pages and classify such accesses across device classes.

[13]Interestingly, there is a great deal of information available on today's mobile handsets that is rarely exploited (such as battery levels, device status, etc.). Novel future services may exploit this information (see Chapter 6).

enable dynamic configuration of services across these environments. This requires breakthrough innovations in:

- Understanding and exploitation of user settings and mobile device configuration and capabilities;
- Service architecting and mobile user experience methods;
- Policy-based decision making and embedded/small footprint policy engine implementations;
- Large-scale and real-time content-aware event processing.

Much research is being done in both industry and academia on these fronts, and can be seen regularly in research periodicals. The market is also seeing some distinct forward-looking application areas into which clusters of services are appearing.

8.4 SUMMARY

This chapter has illustrated, using examples from historical, current and future innovative technologies, that mobile services are an incredibly rich and profound area. Although predictions as to how future mobile services will look are hard to make, the temptation to guess at the next big thing is almost too great to resist. It seems relatively safe to say that there are lots of indication that, in addition to the services we discussed in this chapter and throughout the book, services related to mobile entertainment and mobile social networking will be prominent. Generalizing, there are two main things that humans inherently enjoy:

- Seeking pleasurable and enjoyable experiences
- Communicating with others

Context-aware mobile gaming takes players from fixed consoles out into the real world, which would seem to be a big step in the right direction. Many location-enhanced games have already been developed around the world. Their widespread adoption depends on carriers having the appropriate network and pricing structures in place. Furthermore, context-aware communities that bring together people who share common needs or desires such as social communities, political communities, and gaming communities will also be big and we have already seen signs of this in location-enabled mobile friend finders and messaging services. There is no end to the human need to form communities with like-minded individuals; the only aspect that changes is the technology that facilitates the building and maintaining of those communities. There is also no end to our desire to explore and experience our world and the accumlated knowledge of the world's people. Inasmuch as mobile services will allow that to happen more fluidly, cheaply, and to an ever-widening swath of the Earth's people,[14] we should optimistically expect great things.

[14]See One Laptop per Child, http://laptop.org

QUESTIONS

1. How do you define whether a mobile service is a "success"? What are the dimensions? Try to put yourself in at least three different roles and rethink the question. For example, how does a venture capitalist define success? What about an Internet Search startup? What about a teenage open-source developer? What about the CEO of a nonprofit? What are the intangible factors that may make future mobile services a success? Marketing? Media?

2. Applications such as those developed for the MySpace Web platform are transient (e.g., they may not be posted forever) but can be wildly viral and popular. Some get more usage "hits" and attention than services built by large formal traditional brick-and-mortar companies and organizations. What is the future of mobile services? Platform-centric applications like on MySpace, open-source services, or traditional ones generated by verticals and telecom operators and service providers?

9

CONCLUDING REMARKS

Today, the vision of intelligent mobile services that adapt to user activities, needs, and schedules is becoming a reality; innovative new mobile services are entering the marketplace at an accelerated pace. From location-sensitive information services to social networking to mobile automotive telematics and enhanced reality mobile applications, the richness of new offerings signifies a new era in technology as it starts to serve the needs of individuals and society. Although the decades' old vision is now materializing, we are still facing a number of fundamental technical challenges in several software areas that need to be understood in order for mobile services to achieve the scale, scope, and agility required to make them both relevant and successful. Business challenges also abound in this new world of mobile applications and services. A delicate balance amongst business models, service features and functionality, and price points is required for achieving wide user adoption and leading to a sustainable business.

At the introduction to his book *Smart Mobs*, Howard Rheingold says, "The first sign of the next shift began to reveal itself to me on a spring afternoon in the year 2000. That was when I began to notice people on the streets of Tokyo staring at their mobile phones instead of talking to them. The sight of this behavior, now commonplace in much of the world, triggered a sensation I had experienced a few times before – the instant recognition that a technology is going to change my life in ways I can scarcely imagine" [RHE]. How does one develop the "eye" to see big changes coming? You have to understand their essence. We are seeing more and more of these disruptive behavioral and technological changes.

The Fabric of Mobile Services: Software Paradigms and Business Demands, by Shoshana Loeb, Benjamin Falchuk, and Euthimios Panagos

An integral part of the transformative effect of technology today is the mobile device; it not only gives us the freedom to pursue our personal agenda, but is highly connected and connects us to "a crowd" of our choice. This polarization of extreme personal pursuit combined with extreme connectivity where every move and thought can be shared in real time, is very unique to our times.

In this book, had we attempted to take a detailed snapshot of the mobile services market today, we would have produced a laundry list of technologies, business events and speculations that may have ended up obsolete before the book saw first press. If you browse the daily technology and business blogs, you can see what we mean – it is hard to make sense of all that is happening and sort through what appear to be opposing actions, opinions, and customer realities. One of the things that this book has tried to accomplish is to present the interwoven threads that comprise mobile services and to examine some of them in considerable technical detail. In these technical examples, we hope that we have presented the essence of some key issues in mobile services, issues that will stand the test of time. But the book has also stepped back to take a big look at how the mobile services market is changing and how mobile services both thrive in ones we know and also create totally new ones.

We are all a part of the mobile services fabric, from users to developers, front end to back end. So, what will you weave?

REFERENCES

[ABO] G.D. Abowd, E.D. Mynatt. "Charting Past, Present, and Future Research in Ubiquitous Computing." *ACM Transactions on Computer-Human Interaction*, Vol. 7, No. 1, March 2000, pp. 29–58.

[AKA] Akamai Technologies. "Fast Internet Content Delivery with FreeFlow." White Paper, Boston, MA, 1999.

[ALC] A. Alchian. "Uncertainty, Evolution and Economic Theory." *Journal of Political Economy*, Vol. 58, pp. 211–221, 1950.

[ALE] C. Alexander, S. Ishikawa, M. Silverstein. *A Pattern Language: Towns, Buildings, Construction*. Oxford University Press, 1977.

[AMD] G. Amdahl. "Validity of the Single Processor Approach to Achieving Large Scale Computing Capabilities." *AFIPS Conf. Proceedings* 30: 483–484. April 18–20. Atlantic City, NJ, 1968.

[AND] C. Andersson, D. Freeman, I. James, A. Johnston, S. Ljung. *Mobile Media and Applications – From Concept to Cash*. John Wiley and Sons, 2006.

[AND1] C. Anderson. *The Long Tail: Why Is the Future of Business Selling Less for More*. Hyperion, 2006.

[AND2] C. Anderson, D. Freeman, I. James, A. Johnson, S. Ljung. *Mobile Media and Applications: From Concept to Cash*. Wiley, 2006.

[AND3] C. Anderson. "Free! Why $0.00 Is the Future of Business." *Wired Magazine*, February 2008.

[ANE] N. Anerousis, E. Panagos. "Making Voice Knowledge Pervasive." *IEEE Pervasive Computing*, April–June 2002.

[ARI] D. Ariely. *Predictably Irrational*. Harper Collins, 2008.

[ASH] R. Ashkenas. "Simplicity-Minded Management." *Harvard Business Review*, 85(12), pp. 101–110, December 2007.

[BAL] C.Y. Baldwin, K.B. Clark. *Design Rules: The Power of Modularity*. The MIT Press, 2000.

[BALD] M. Baldauf, S. Dustdar, F. Rosenberg. "A Survey on Context-Aware Systems." Technical Report TUV-1841-2004-24, Technical University of Vienna, 2007.

The Fabric of Mobile Services: Software Paradigms and Business Demands, by Shoshana Loeb, Benjamin Falchuk, and Euthimios Panagos

[BAR] J.R. Barry. "Wireless Infrared Communications." *The Springer International Series in Engineering and Computer Science*. August 31, 1994.

[BARL] R. Bartle. "Mud, Mud, Glorious Mud." An article on Multi-User Dungeon written by one of the game's authors. http://www.mud.co.uk/richard/masep84.htm.

[BART] N. Bartolini, E. Casalicchio, S. Tucci. "A Walk through Content Delivery Networks." Tutorial in MASCOTS 2003. Pages 1–25.

[BAT] J. Battelle. "The Search: How Google and Its Rivals Rewrote the Rules of Business and Transformed Our Culture." *Portfolio*, 2005.

[BEN] Y. Benkler. *The Wealth of Networks; How Social Production Transforms Markets and Freedom*. Yale University Press, 2006.

[BER] T. Berners-Lee, J. Hendler, O. Lassila. "The Semantic Web: A New Form of Web Content That Is Meaningful to Computers Will Unleash a Revolution of New Possibilities." *Scientific American Magazine*, May 2001.

[BIL] A. Biliris, E. Panagos. "A High Performance Configurable Storage Manager." IEEE 11th Int'l Conference on Data Engineering, Taipei, Taiwan. March 1995.

[BRA] A. Brandenburger, B. Nalebuff. "Co-Opetition: A Revolution Mindset That Combines Competition and Cooperation." Doubleday, 1996.

[BRU] G.C. Bruner, A. Kumar. "Attitudes Towards Location Based Advertising." *Journal of Interactive Advertising*, Vol. 7, No. 2, 2007. (http://www.jiad.org/article89).

[BUR] B. Burnham. "Why the Flow of Innovation Has Reversed." Union Square Ventures Blog, http://www.unionsquareventures.com, September 2008.

[CAM] G. Camarillo, M. Garcia-Martin. *The 3G Multimedia Subsystem (IMS): Merging the Internet and the Cellular Worlds*, 3rd ed. Wiley, 2008.

[CAR] N. Carr. "Is Google Making Us Stupid?" TheAtlantic.com, http://www.theatlantic.com/doc/200807/google, July/August 2008.

[CAR2] N. Carr. *The Big Switch: Rewiring the World, from Edison to Google*. W.W. Norton and Company, 2008.

[CHA] B.L. Chalfonte, R.S. Fish, R.E. Kraut. "Expressive Richness: A Comparison of Speech and Text as Media for Revision." In *Proceedings of the Conference on Computer Human Interaction*, pages 21–26. ACM, April 1991.

[CHO] H.D. Chon, S. Jun, H. Jung, S.W. An. "Using RFID for Accurate Positioning." In *Journal of Global Positioning Systems*, 2004. Vol. 3, No. 1–2:32–39.

[CON] D. Conley. *Elsewhere, U.S.A.* Pantheon Books, 2009.

[CUL] B. Cully, G. Lefebvre, D. Mayer, M. Feeley, N. Hutchinson. "Remus: High Availability via Asynchronous Virtual Machine Replication." 5th USENIX Symposium on Networked Systems Design and Implementation, April 16–18, 2008, San Francisco, CA.

[DAN] R. Dantas, J. Fidalgo, D. Sadok, C. Kamienski, B. Ohlman. "Policies for the Management of Ambient Networks: From Theory to Practice." *Proc. IEEE Workshop on Policies for Distributed Systems and Networks*, pp. 37–45, June 2008.

[FAL2] B. Falchuk, D. Shallcross. "Simulating and Implementing Geospatially-Based Binding Mechanisms for Mobile Peering." *Proc. of ICST/CREATE-NET Int'l Conf. on Ambient Media and Systems* (Ambi-sys 2008), Quebec City, 2008.

[FAL3] B. Falchuk, S. Loeb, M. Lai, G. Cohen. "System for Controlling Mobile Device Behavior in Multi-Network Areas." *Proc. IEEE Consumer Communications & Networking* (CCNC 2007), Las Vegas, 2007.

[FAL4] B. Falchuk, S. Loeb, D. Famolari. "Location-Based Services." In *Encyclopedia of Wireless and Mobile Communications*, B. Furht (ed.), CRC Press, 2007.

[FAL5] B. Falchuk, S. Loeb, T. Panagos. "Seamless Mobile Context." *Proc. IEEE Consumer Communications and Networking Conference* (CCNC), Las Vegas, 2008.

[FAL08] B. Falchuk, S. Loeb, E. Panagos. "A Deep-Context Personal Navigation System." In *15th World Congress on ITS*, New York. November 2008.

[FIN] C.H. Fine. *Clockspeed: Winning Industry Control in the Age of Temporary Advantage*. Basic Books, 1998.

[FIS] C.S. Fischer. *America Calling: A Social History of the Telephone to 1940*. University of California Press, 1992.

[FISH] P. Fishwick (ed.). *Aesthetic Computing*. MIT Press, 2006.

[FRE] S. Freud. *Civilization and Its Discontent*. W.W. Norton Company, 2005.

[FUR] G. Furnas. "Generalized Fisheye Views. Proc. of Human Factors in Computing Systems CHI '86, Boston, April 13–17, 1986, pages 16–23.

[GAM] E. Gamma, R. Helm, R. Johnson, J.M. Vlissides. *Design Patterns: Elements of Reusable Object-Oriented Software*. Addison-Wesley Professional Computing Series, 1994.

[GLA] M. Gladwell. *The Tipping Point*. Backbay, 2002. (Hardcover published by Little Brown and Company in 2000.)

[GLA1] M. Gladwell. *Blink: The Power of Thinking Without Thinking*. Back Bay Books, 2007.

[GOL] P. Golding. *Next Generation Wireless Applications*. 2nd ed. Wiley, 2008.

[GON] C. Gonsalves. "Multiple Online Personas: The Choice of a New Generation." *Baseline Magazine*, http://www.baselinemag.com, February 8, 2008.

[GRE] A. Greenfield. "Everware: The Dawning Age of Ubiquitous Computing." New Riders, Voices that Matter. 2006.

[GRO] A.S. Grove. *Only the Paranoid Survive: How to Exploit the Crisis Points That Challenge Every Company*. Doubleday Business, 1999.

[GRU] M. Gruteser, D. Grunwald. "Anonymous Usage of Location-Based Services Through Spatial and Temporal Cloaking." *Proceedings of the 1st International Conference on Mobile Systems, Applications and Services*. San Francisco, 2003.

[GRUB] R.E. Gruber, B. Krishnamurthy, E. Panagos. "CORBA Notification Service: Design Challenges and Scalable Solutions." IEEE 17th Int'l Conference on Data Engineering, Heidelberg, Germany, April 2001.

[GUN] N.J. Gunther. *The Practical Performance Analyst*. iUniverse Press, 2000.

[HAL] J. Hallberg, M. Milsson. "Positioning with Bluetooth, IrDA and RFID." Master's Thesis. Lulea University of Technology. Department of Computer Science and Electrical Engineering. 2002.

[HEL] M. Heller. *Gridlock Economy*. Basic Books, 2008.

[HIN] R. Hinman. "Ubiquitous Computing Workshop: Mobile User Experience Design Principles." http://www.adaptivepath.com/blog/2007/09/17/ubiquitous-computing-workshop-mobile-user-experience-design-principles/

[HIND] D. Hindus, C. Schmandt, C. Horner. "Capturing, Structuring and Representing Ubiquitous Audio." *ACM TOIS, Transactions on Office Systems*, October 1993.

[HOR] E. Horvitz, P. Koch. "Mobile Opportunistic Planning: Methods and Models." *Proc. Conf. on User Modeling*, June 2007, Corfu, Greece.

[HUB] D.W. Hubbard. *How to Measure Anything*. John Wiley and Sons, 2007.

[HUG] M. Hughes. "Buzzmarketing: Get People to Talk about Your Stuff." *Portfolio*, 2005.

[IAN] M. Iansiti, R. Levien. *The Keystone Advantage: What the New Dynamics of Business Ecosystems Mean for Strategy, Innovation, and Sustainability*. Harvard Business School Press, 2004.

[IBM07] IBM Institute for Business Value Analysis Based on Deutsche Bank Global Markets Research on Nokia, October 19, 2007.

[IBM08] IBM Institute for Business Value Study. "Go mobile, grow ... Should mobile Internet services be the next big growth gamble for mobile device makers?" Last updated May 30, 2008.

[IDC08] IDC. "Worldwide Mobile Phone 2008–2012 Forecast and Analysis: Continued Growth." Document #211441, March 2008.

[INT06] Intel Corporation and NTT DoCoMo. "Open and Secure Terminal Initiative (OSTI) Architecture Specification, Revision 1.00." 2006.

[JAN] B. Janssen. "The Kindle and the iPhone Dance." *Publishing Frontier*, July 2008. http://pubfrontier.com/2008/07/20/e-ink-the-kindle-and-the-iphone/

[JUN] J. Jung, B. Krishnamurthy, M. Rabinovich. "Flash crows and denial of service attacks: Characterizations and implications for CDNs and Web sites." WWW 2002. Pages: 293–304.

[KUP] A. Kupper. *Location-Based Services: Fundamentals and Operation*. John Wiley and Sons, England, 2005.

[LAM] M. Lamming, M. Flynn. "Forget-me-not" Intimate Computing in Support of Human Memory." *Proceedings of FRIEND21, '94 International Symposium on Next Generation Human Interface*, 2–4 February 1994, Meguro Gajoen, Japan.

[LAW] G. Lawton. "US Cell Phone Industry Faces an Open Future." *IEEE Computer*, 41(2), pp. 15–18, February 2008.

[LOE] A. Loeb, T.J. Cox. "Our Galaxy's date with Destruction." In *Our Galaxy's Collision with Andromeda*, Astronomy, June 2008.

[LUC] R. Lucky. "The Illusive Future." *IEEE Spectrum*, September 2008.

[MADE] J. Maeda. *The Laws of Simplicity*. MIT Press, 2006.

[MAE] P. Maes. "Agents that reduce work and information overload." *Communications of the ACM*, pp. 30–40, 37(7), 1994.

[MCC] T. McCourt, C. O'Donnell, J. Patterson. "What happens when wireless truly becomes the next computing platform?" Morgan Keegan Analyst Report, 2007.

[MER] D. Merrill. "Mashups: The new breed of Web App." Available on IBM DeveloperWorks, http://www.ibm.com/developerworks, 2006.

[MIC] B. Michelson. "Event-Driven Architecture Overview." Seybold Group whitepaper, available at http://www.psgroup.com.

[MOD] B. Modoff, J. Goldberg. "Shifting Sands: Disruptive Wireless Market Drivers, Trends and Catalysts." Global Market Reseach, Deutsche Bank Securities, Inc., 2008.

[MOG] B. Moggridge. *Designing Interactions.* MIT Press, 2007.

[MOR] P. Morville. *Ambient Findability.* O'Reilly Media, 2005.

[MUL] *Intelligence in Services and Networks: Technology for Cooperative Competition.* Edited by A. Mullery, M. Besson, M. Campolargo, R. Gobbi, R. Reed. Lecture Notes in Computer Science, Springer-Verlag, Berlin, 1997.

[MUR] P. Murphy, E. Welsh, J.P. Frantz. "Using Bluetooth for Short-Term Ad Hoc Connections Between Moving Vehicles: A Feasibility Study." In *Vehicular Technology Conference, 2002*, VTC Sprint 2002, IEEE 55th.

[NAF] "Just How Context Aware Are You? Mobility, Datastreams, and What Inferencing Does and Doesn't Make Possible." Keynote speech at the Web 2.0 conference, New York, September 2008.

[NAI] J. Naisbitt. *Megatrends.* Warner Books, 1982.

[NIS] *Algorithmic Game Theory.* Edited by Noam Nisan, Tim Roughgarden, Eva Tardos and ViJay V. Vazirani, Cambridge University Press, 2007.

[NOR] D.A. Norman. *The Design of Everyday Things.* Basic Books, 1988.

[NOR2] D.A. Norman. *Emotional Design.* Basic Books, 2004.

[NOT] Object Management Group. "Notification Service Specification Version 1.1." October 2004.

[ORE] T. Oreilly. "What Is Web 2.0." 2005. http://www.oreillynet.com/pub/a/oreilly/tim/news/2005/09/30/what-is-web-20.html

[ORI] T. O'Reilly. O'Reilly Radar, http://radar.oreilly.com/tim/

[PAN] E. Panagos, A. Delis. "Selective Replication for Content Management Environments." *IEEE Internet Computing*, Vol. 9, No. 3, May–June 2005.

[PIN] P.J. Pine, J.H. Gilmore. *The Experience Economy: Work Is Theater and Every Business a Stage.* Harvard Business School Press, 1999.

[POU] Z. Pousman, J. Stasko. "A Taxonomy of Ambient Information Systems: Four Patterns of Design." *Proc. ACM AVI'06*, Venezia, May, 2006.

[QLV] Q. Lv, P. Cao, E. Cohen, K. Li, S. Shenker. "Search and Replication in Unstructured Peer-to-Peer Networks." In *Proceedings of the 16th ACM International Conference on Supercomputing*, New York, June 2002.

[RAP] J. Raper, G. Gartner, H. Karimi, C. Rizos. "Applications of Location-Based Services: A Selected Review." *Journal of Location Based Services*, 1(2), June 2007.

[RAV] N. Ravi. "Context-Aware Battery Management for Mobile Phones: A Feasibility Study." Microsoft Research Tech Report, available at http://reality.media.mit.edu/publications.php, 2006.

[RAY] J. Rayport. "The Virus of Marketing." *Fast Company*, December 1996.

[RES] P. Resnick. "HyperVoice: A Phone-Based CSCW Platform." In *Computer Supported Cooperative Work—CSCW'92 Conference Proceedings.* ACM, New York, pp. 218–225, 1992.

[RHE] H. Rheingold. *Smart Mob: The Next Social Revolution.* Basic Books, 2002.

[RIG] P. Rigaux, M. Scholl, A. Voisard. *Spatial Databases with Applications to GIS*. Morgan Kaufmann, 2002.

[SCH] B.N. Schilit, N. Adams, R. Want. "Context Aware Computing Applications." *IEEE Workshop on Mobile Computing Systems and Applications*. December 1994.

[SCHM] C. Schmandt, B. Arons. "Phone Slave: A Graphical Telecommunications Interface." In *Proceeding of the Society for Information Display (SID)*, Vol. 26/1, 1985.

[SCO] R. Scoble, S. Israel. *Naked Conversations: How Blogs Are Changing the Way Businesses Talk*. Wiley, 2006.

[SHA] E. Shanahan, P. Thurneyson. "Quality of Experience Drives Mobile Content." available at http://www.billingworld.com, posted March, 2007.

[SHN] B. Shneiderman. *Leonardo's Laptop: Human Needs and the New Computing Technologies*. MIT Press, 2002.

[SIL] G. Silverman. "The Secrets of Word-of-Mouth Marketing: How to Trigger Exponential Sales Through Runaway Word of Mouth." Amacom 2001.

[STA] T. Standage. *The Victorian Internet*. Walker and Company, 1998.

[STU] Studio 7.5. *Designing for Small Screens*. Ava Publishing, Switzerland, 2005.

[SUB] S.R. Subramanya, B.K. Yi. "Enhancing the User Experience in Mobile Phones." *IEEE Computer*, 40(12), pp. 114–117, December 2007.

[SUR] J. Surowiecki. *The Wisdom of the Crowd*. Doubleday, 2004.

[TGR] T. Gruber. "A Translation Approach to Portable Ontology Specifications." *Knowledge Acquisition*, 5(2):199–220, 1993.

[THE] S. Androutsellis-Theotokis, D. Spinellis. "A Survey of Peer-to-Peer Content Distribution Technologies." *ACM Computing Surveys* (CSUR), 36(4), Pages: 335–371, December 2004.

[TOL] N. Tolia, D. Andersen, M. Satyamarayanan. "Quantifying Interactive User Experience on Thin Clients." *IEEE Computer*, pp. 46–52, March 2006.

[TUL] E. Tulving. *Elements of Episodic Memory*. Clarendon Press, 1983.

[TUR] B. Turner. "Service Creation – Traditional Telecom Still Doesn't Get It." *Internet Telephony Magazine*, April 2008.

[VER] R. Vertegaal, I. Poupyrev. "Organic User Interfaces." *CACM*, 51(6), pp. 26–29, June 2008.

[VXML] World Wide Web Consortium (W3C), "Voice Extensible Markup Language (VoiceXML) Version 2.0."

[W3C] W3C. "Mobile Web Best Practices 1.0, Basic Guidelines" W3C Proposed Recommendation, 2 November 2006.

[WAT] D.J. Watts. "A Twenty-First Century Science." *Nature*, Vol. 445, February 2007, p. 489.

[WEI] M. Weiser. "The Computer for the 21st Century." *Scientific American Special Issue on Communications, Computers, and Networks*, September 1991.

[WEI2] M. Weiser, R. Gold, J.S. Brown. "The Origins of Ubiquitous Computing Research at PARC in the Late 1980s." *IBM Systems Journal*, 38(4), 1999.

[WEI3] M. Weiser, J. Brown. "Designing Calm Technology." *PowerGrid Journal*, v 1.01, 2001.

[WIG] D. Wigdor, C. Forlines, P. Baudisch, J. Barnwell, C. Shen. "LucidTouch: A See-Through Mobile Device." *Proc. ACM UIST'07*, Rhode Island, 2007.

[WOO] M. Wooldridge, N. Jennings. "Intelligent Agents: Theory and Practice." *The Knowledge Engineering Review*, 10(2), 115–152, 1995.

[YAN] S. Yang. "Engineers Harness Cell Phone Technology for Use in Medical Imaging." *UC Berkeley ScienceDaily*, May 1, 2008.

[YAT] S. Yates. "Worldwide PC Adoption Forecast 2007 To 2015." Forrester Research, June 2007.

[YIN] L. Yin, G. Cao, C. Das, A. Ashraf. "Power-Aware Prefetch in Mobile Environments." *22nd IEEE Int'l Conference on Distributed Computing Systems*, Vienna, Austria. July 2002.

INDEX

The Fabric of Mobile Services Software Paradigms and Business Demands, by Shoshana Loeb, Benjamin Falchuk, and Euthimios Panagos